HISTORY OF INDIAN AIR DEFENCE ARTILLERY

ABOUT THE UNITED SERVICE INSTITUTION OF INDIA

Established in 1870 by Major General Sir Charles MacGregor in Shimla, the United Service Institution of India (USI) is a national security and defence services think tank based in New Delhi, India. Its aim is the 'furtherance of interest and knowledge in the art, science and literature of the defence services.' It is located at Vasant Vihar, New Delhi, from where it has been operating since 1996.

ABOUT THE CENTRE FOR MILITARY HISTORY AND CONFLICT STUDIES

(formerly Centre for Armed Forces Historical Research)

The Centre was established in December 2000 under the aegis of the USI for encouraging the objective study of all facets of Indian military history with a special emphasis on the history of the Indian Armed Forces. It focuses on diverse aspects of the history of Indian military evolution, policies and practices—strategic, tactical, logistical, organisational and socio-economic, as well as the field of contemporary conflict studies.

HISTORY OF INDIAN AIR DEFENCE ARTILLERY 1940-1945

Colonel Mandeep Singh (Retired)

CENTRE FOR MILITARY HISTORY AND CONFLICT STUDIES
UNITED SERVICE INSTITUTION OF INDIA

MANOHAR
2024

First published 2022

Reprinted 2024

ISBN 978-93-91928-64-3

Published by
Ajay Jain *for*
Manohar Publishers & Distributors
4753/23 Ansari Road, Daryaganj
New Delhi 110002

Typeset by
Ravi Shanker
Delhi 110095

Printed at
Replika Press Pvt. Ltd.

Contents

List of Illustrations 7

Acknowledgments 9

Prologue 11

Abbreviations 13

Introduction 17

1. A Beginning is Made: The First Raisings 29
2. Hong Kong-Singapore Royal Artillery: The First Experience with Anti-Aircraft Artillery 39
3. Baptism: The Campaign in Malaya, 1941-1942 46
4. Retreat from Burma, 1941-1942 72
5. Defending the Homeland 93
6. Prelude to the Return to Burma 125
7. Back to Burma 135
8. Other Theatres and Lesser Campaigns 157
9. After the War 167

Conclusion 176

Epilogue 189

Appendices

1. Japanese Air Forces and Air Tactics 193
2. Anti-Aircraft Guns 198

3. Light AA/Anti-Tank Regiments 201
4. Raising of Indian Air Defence Units 205
5. Composition of Indian AA Regiments 1940-1945 212
6. Indian AA Order of Battle 215
7. Honours and Awards 221
8. Disbandment of AA Units 225
Bibliography 227
Index 235

Illustrations

MAPS

1. AA Defences in Singapore 47
2. Burma 1941-2 73
3. Map showing deployment of AA Brigades and Regiments, April 1943 94
4. Allotment of AA Resources for 14th Army Offensive 1943-4 136

PLATES* *(between pp. 120-1)*

1. Madrassi LAA gunners
2. LAA Battery of 25 Infantry Division at Maungdaw, September 1945
3. 19 HAA Battery of 7 Rajput HAA Regiment in Assam, February 1945
4. Madras AA gunners defending the No 1 Squadron, RIAF airfield
5. Shot down Japanese Bombers on public display in Calcutta, January 1943
6. Commander-in-Chief visits LAA site Calcutta, November 1944
7. Commander-in-Chief visits Maratha LAA Battery at Kohat, December 1944
8. Maharaja of Faridkot visits LAA site Karachi, January 1944
9. No. 3 AATC, Mehgaon, 1943
10. 16 HAA Battery/6 Indian HAA Regiment, RIA, 1944

*Photographs collected from Armed Forces Films and Photo Division, H-Block, Ministry of Defence, New Delhi.

11. LAA Training of recruits at FATC, Mathura, 1943
12. 15 Indian LAA Regiment at Malir, Cantt, October 1944
13. AA Defences, Bombay Port, 1944
14. LAA Battery undergoing Combined Operations Training at Combined Training Centre, Bombay, April 1944
15. 19 HAA Battery of 7 Rajput HAA Regiment in Arakan, December 1944
16. 6 Punjab LAA Regiment Contingent at the Delhi Victory Parade, March 1946

Acknowledgements

It is often said that writing a book is a lonely and tiresome process but I have been fortunate to have received the support and encouragement of a number of friends and colleagues while writing this book. It has never been my effort alone but that of my family, friends and colleagues. I am grateful to all of them for without them, this work would not have been possible.

At the outset, I wish to acknowledge with thanks the United Service Institution of India Centre for Military History and Conflict Studies (CMHCS) for giving me the opportunity to carry out the study and research on the History of Indian Air Defence Artillery 1940-45 under the aegis of the Chhatrapati Shivaji Chair Fellowship (2019-20).

I gratefully acknowledge with thanks Lieutenant General P.K. Singh, PVSM, AVSM, former Director, USI, Major General B.K. Sharma, AVSM, SM, Director, USI, and Major General P.K. Goswami, VSM, for their support and guidance.

I am deeply indebted to Major General Ian Cardozo, AVSM, SM, Chairman, CMHCS and Squadron Leader Rana T.S. Chhina, MBE, Secretary, CMHCS, for their encouragement, constant support and advice while carrying out this research project and writing of the book.

A special word of thanks to Lieutenant General A.P. Singh, PhD, Director General, Army Air Defence, for his encouragement and advice that immensely helped me in completing this project.

The records at the History Division and Armed Forces Films and Photo Division, Ministry of Defence, have been of immense help in carrying out the study and research on the subject. For providing access to the records and helping me in the study of the same, I am thankful to Shri Om Prakash, former Director, and Dr Narendra Yadav, Deputy Director of History Division, Ministry of Defence, Government of India, and Shri N. Shrikumar, Director, Armed Forces Films and Photo Division, Ministry of Defence, Government of India, for without their support this project could not have been completed.

I embarked on this journey only because of the encouragement and support of Lieutenant Colonel A.S. Bajwa (Baaz) and it is he who has kept me going on. For this, I will be forever indebted to him. A special thanks to Baaz, whom I am proud to call my friend.

Last but not the least, my thanks and gratitude to my wife Harpreet and my daughters, Prabhleen and Gurleen, for all the sacrifice, support and help that they have given me over the years, all the while I have been busy with this book. It could not have been possible without them.

COLONEL MANDEEP SINGH (Retd.)

Prologue

On 12 December 1971, No. 28 Squadron, Indian Air Force carried out a strike against Tezgaon by a four-ship formation of MiG-21s armed with bombs and led by Squadron Leader K.J.S. Gill. The deadline for the expiry of the ceasefire given by India loomed and Gill wanted to carry out one last strike against the main Pakistan Air Force base in (then) East Pakistan. At 2:00 p.m., Gill's MiG-21 took off from Guwahati with Flight Lieutenants N.S. Malhi, Manbir Singh and D.M. Subaiya.

The MiGs reached Tezgaon without any incident and as they attacked the already damaged airfield, the 500 lb. bombs found their mark and scored direct hits. The northern part of the runway was put totally out of commission.

The airfield was defended by a battery of 37mm. anti-aircraft guns of the 6th Light AA Regiment of Pakistan Army. With no air force to bank on, the anti-aircraft battery had been hard pressed to counter the constant and relentless raids by the Indian Air Force. Its ammunition stock had depleted to dangerously low levels by continuous firing since the start of the war. With no chance of any resupply or an end to the Indian air raids, Lieutenant Colonel Mohammad Afzal, the Commanding Officer of the 6th Light, had already issued orders that the guns would fire only three rounds per engagement to preserve the ammunition. This was, of course, not known to the Indians.

As the MiGs were carrying out the raid, the anti-aircraft guns opened up and they also found their mark. Gill's MiG-21 was hit. The aircraft started streaming smoke and as Gill pulled away, he called out that he had been hit. His wingman, Flight Lieutenant Malhi followed him as Gill turned towards Agartala, the nearest base. Malhi mistakenly engaged reheat and in the process overshot Gill, and lost contact.

Gill nursed his damaged aircraft as long as possible but as he neared the Indian border, the fire had spread, forcing Gill to eject. He was picked up by Mukti Bahini who transported him to Agartala in the night.

Gill's MiG had been shot down by an anti-aircraft gun of 6th Light AA Regiment, firing just three rounds. This regiment was originally the first Indian Light AA Regiment, raised in 1941, and renumbered twice in its service. In 1947, the Regiment was transferred to Pakistan but its lineage can be traced back to origin of Indian Air Defence Artillery. This book is about the same: *The History of Indian Air Defence Artillery, 1940-1945.*

Abbreviations

AD	Air Defence
AMES	Air Ministry Experimental Stations
ARP	Air Raid Precautions
AVG	American Volunteer Group
ALFSEA	Allied Land Forces South East Asia
AA	Anti-Aircraft
AACU	Anti-Aircraft Cooperation Unit
AADC	Anti-Aircraft Defence Commander
AAOR	Anti-Aircraft Operation Room
AATC	Anti-Aircraft Training Centre
A/Tk	Anti-Tank
Bty	Battery
BC	Battery Commander
Bde	Brigade
BAD	Brigadier Air Defence
BAA	Brigadier Anti-Aircraft
BRA	Brigadier Royal Artillery
BOR	British Other Ranks
BAF	Burma Auxiliary Force
Cantt	Cantonment
Capt	Captain
COD	Central Ordnance Depot
CGA	Ceylon Garrison Artillery
COL	Chain Overseas Low
Col	Colonel
CRA	Commander Royal Artillery
CO	Commanding Officer
Div	Division
EW	Early Warning

EW	Electronic Warfare
GHQ	General Headquarters
GOC	General Officer Commanding
GCI	Ground Control Interception
GL	Gun Laying
GOR	Gun Operations Room
GPO	Gun Position Officer
GPOA	Gun Position Officer-Assistant
Hy	Heavy
HAA	Heavy Anti-Aircraft
in.	Inch
IDSM	Indian Distinguished Service Medal
IOM	Indian Order of Merit
IOR	Indian Other Ranks
IJAAS	Imperial Japanese Army Air Service
IJN	Imperial Japanese Navy
JCO	Junior Commissioned Officer
KCIO	King's Commissioned Indian Officer
Lt	Lieutenant
Lt Col	Lieutenant Colonel
Lt Gen	Lieutenant General
Lt	Light
LAA	Light Anti-Aircraft
Maj	Major
Maj Gen	Major General
MGRA	Major General Royal Artillery
Mk	Mark
MT	Mechanised Transport
mm	Millimetre
MC	Military Cross
MES	Military Engineering Service
MM	Military Medal
MATC	Mountain Artillery Training Centre
NCO	Non-Commissioned Officer
No.	Number
OC	Officer Commanding
Orbat	Order of Battle

OBE	Order of the British Empire
OBI	Order of British India
OR	Other Ranks
PAD	Passive Air Defence
lb.	Pound
PM	Punjabi Mussalman
QVO	Queen Victoria's Own
Rdr	Radar
RDF	Radio Direction Finding
Regt	Regiment
RH	Road-head
RAF	Royal Air Force
RA	Royal Artillery
RATC	Royal Artillery Training Centre
RIA	Royal Indian Artillery
RIAF	Royal Indian Air Force
RGA	Royal Garrison Artillery
2 ic	Second in Command
SP	Self Propelled
Tp	Troop
UK	United Kingdom
VCO	Viceroy Commissioned Officer
VA	Vulnerable Area
VP	Vulnerable Point

Introduction

Having made its debut in the Franco-Prussian War of 1870, anti-aircraft artillery played only a minor role in the war and had not yet matured when World War I ended. A major reason for this was that developments in the field of air power notwithstanding, it had been a land centric war. Air power may have changed the nature of warfare but it was yet to play a decisive role in its conduct. With limited impact and threat from the air, anti-aircraft artillery was never a major player during the war.

It was thus no surprise that one of the largest cutbacks by all nations after the Great War were in anti-aircraft artillery and the United Kingdom was no exception. At the end of the Great War, Royal Artillery had 48 AA Companies of Royal Garrison Artillery (RGA), 31 anti-aircraft battery HQs, 225 anti-aircraft sections and three mobile anti-aircraft brigades. By the end of 1919, the entire force of regulars, conscripts, reservists and naval auxiliaries committed to anti-aircraft defences had been broken up, leaving only one regular anti-aircraft brigade of three gun anti-aircraft batteries and one Royal Engineers searchlight battalion.[1]

In the 1920s, the realisation that air could be used against British imperial interests and installations around the globe led to a gradual increase in the allocation of AA defences for overseas British ports and bases. Anti-Aircraft brigades and batteries were once again raised to meet the overseas commitments and were deployed at the major ports and bases, including Singapore, Hong Kong and Malta. Only one such anti-aircraft battery was not deployed at any port and was located in India for defence against an overland offensive. The AA battery was deployed in the North West Frontier Province in India.[2]

In March 1921, the Legislative Assembly of India had passed

a resolution defining the role of Army in India as 'the defence of India against external aggression and the maintenance of internal peace and tranquility'. For any other purpose, the obligations resting on India were to be optional and self-imposed, as they were in the case of self-governing dominions.[3]

This role of the Army was accepted by the Government of India and His Majesty's Government in the United Kingdom. There were serious implications of accepting this role and though they may be obvious in hindsight, there were, in all probability, not appreciated at that time by the Government of India, for the defence of India against attack by a major power was beyond the capability of the (existing) Army in India. Also, the need of safeguarding its imperial interests and outposts in India would have invariably sucked in resources from the Army in India. There was thus an inter-dependence between India and the United Kingdom and this was to play a pivotal role when the Defence of India plan was being worked out.

With Russia being the only (then) major power that was a threat to British interests in the East, the Defence of India plan was drawn up against an eventuality of an attack by Russia on India, through Afghanistan. The plan envisaged that the Army in India would prevent a Russian invasion by undertaking an advance to halt the Russian forces in Afghanistan. The main drawback of this plan was that the Army in India had only limited capabilities at that stage and could only offer the initial resistance; the main resistance to the purported Russian invasion was to be given by a strong British Expeditionary Force, sent from the United Kingdom. This meant that the publicly accepted role of the Army in India was modified; the change being the first such modification but it was certainly not the last.

While Russia was viewed as a major power capable of invading Afghanistan and threatening India, its Air Force was in no position to pose any serious threat or support its purported invasion with large scale air operations. With such limited air threat, it was appreciated that one AA battery would be an effective deterrent and accordingly the 8th AA Battery arrived in India on 9 November 1926, and came to be located at Peshawar in Northwest India.[4] An independent battery, it was equipped with eight 3-inch 20 cwt. AA guns, organized into a Battery Headquarters with four sections of two anti-aircraft guns each and had a war establishment of 221 all ranks.[5]

The Soviet Air Force was an arm of the Army at the end of World War I and came to be institutionalized only in 1928 as an independent arm in the form of the Forces of Aerial Defence. The Soviet Air Force had a humble start as it had been decimated at the end of the Great War but the close military ties with Germany resulted in not only the transfer of technology but sharing of tactical doctrines in the field of aviation. It was these developments that made the Soviet Air Force a threat on India's North-West Frontier. The threat was real and not something that could be wished away. In 1929, the Soviets had directly intervened when they sent 2,000 troops dressed as locals to support their ally, King Amanullah Khan, who had been deposed by British-supported rebels led by Habibullah Kalakani. The Soviet troops were recalled when Amanullah's forces were defeated near Kabul and he, himself, fled the country. The threat from air may not have been so present but was serious enough to warrant the 8th AA Battery to be deployed in the NWFP. While the gradual expansion of anti-aircraft defences was taking place, Great Britain was also coming to terms with the fact that the end of World War I may have brought in a period of peace but there remained a need to safeguard and defend the widespread imperial interests, including in the Far-East. With its own Army downsized, the War office in London understandably looked at the Army in India as a reserve to be tapped into, as had been done during the Great War when over a million soldiers of the Indian Army fought against the German empire in Europe and Africa. To arrive at a formal and binding agreement, the Director of Military Operations at the War office, in a dispatch dated 23 September 1923, had sought a commitment from the General Staff in India for a desptach of two mixed Indian brigade groups for the defence of the Anglo-Iranian and Mosul oilfields, and a third Indian infantry brigade for garrison duties in the Persian Gulf.[6]

Though the existing resources were already stretched thin in meeting India's own obligations, the Chief of General Staff, India, agreed to meet the 'immediate requirements of the situation' in an emergency. This commitment, though formalized, was not accepted publicly by the General Staff in India.

This change had not even been acknowledged when the Garran Tribunal was constituted to look at the role of the Army in India, and also the associated expenditure. While the major contention was over the tribunal's recommendation that a sum of £1,500,000 be paid annually as grant-in-aid by Great Britain towards the military expenditure from Indian revenue, the recommended role of direct concern was the redefining of the Army's role as 'the protection of the Indian frontiers against local aggression and maintenance of internal security'. From defence against a major power, the role was now limited to 'protection from local aggression'.

For the first time, the overseas commitments were also recognized as a role by the Tribunal and the Army in India was asked to be 'a force, ready in an emergency to take the field at once, which does not exist elsewhere in the Empire, which is specifically available for immediate use in the East and which has on occasions been so used'. As this was against the publicly stated defence policy of India, this role was never accepted and confirmed publicly.

As the discussion and negotiations on the role of the Army in India were going on, the anti-aircraft defences were being consolidated and reorganized by Great Britain. A new anti-aircraft brigade, the 5th AA Brigade, RA, was raised at Hong Kong, with Headquarters and the 7th AA Battery at Peshawar also coming under its command.[7] A related development was the proposal of raising two anti-aircraft batteries of Hong Kong Singapore Royal Artillery for which the recruitment as to be done from India, and more specifically from the Punjab. These were to go on to be the first anti-aircraft batteries manned by Indians and would go on to play an important role in the battles of Hong Kong and Malaya, albeit not forming part of the Indian Army and were Royal Artillery units.

Coming back to the Indian Army, the process on Indianisation of the Army was taking root and, as part of the process, the first Indian artillery unit was raised on 15 January 1935. The process was, however, limited to field artillery and did not envisage any Indian anti-aircraft unit at this stage.

The role of the Army underwent another revision just four years post the Garran Tribunal recommendations and was now defined as 'the defence of her land frontiers against aggression by a second class power; initial operation with available forces and pending the arrival of imperial reinforcements, to protect her land frontiers against aggression by a first class power; the close defence of her coastline against sea-borne aggression; the support of the civil power in the maintenance of internal law and order; and if the situation in India permits, the provision in an emergency of some assistance to the Imperial Forces outside India.'

The formalization of these commitments was not without ambiguity and lack of clarity. While the Government of India had to bear a financial burden of maintaining troops ear-marked for service elsewhere that it could have done without, His Majesty's Government was forced to accept a stipulation that in an emergency it could rely on troops which were to be made available only if the situation in India so permitted. Not satisfied with the existing obligations, the Government of India sought an arrangement vide which the forces required for overseas commitments were to be organized into a separate Imperial Reserve Division and were maintained out of British contributions. The view from London was, however, against placing any conditions for India to meet the overseas commitments.

To come to a more acceptable and binding arrangement, the Chatfield Committee in 1938-9 sought a novel way out. The localities that had till now been treated as imperial outposts were now to be considered as 'bastions of India's External Defence', the defence of which was considered essential for India's defence. Moreover, the responsibility of defending Indian frontiers and the overseas outposts was to be a joint responsibility of the Government of India and His Majesty's Government.[8] With this, the role of the Army in India was amended and the obligation of sending troops overseas became an integral part of the duties of the Army. As a result, the formations meant for duty outside India became an integral part of the Army and were known as 'External Defence Troops'.

From a reluctant acceptance of the limited obligation of sending troops overseas, the Army in India was now bound by an unlimited and unconditional liability for overseas reinforcements. Not only this, the Chatfield Committee expanded the obligation further by adding "that there may be occasions, as in the past, on which the Government of India will be able and willing to make forces available for purposes going beyond the external defence of India".

As part of its commitment to safeguard India's frontiers, the Army had drawn up a number of plans for the North West Frontier, starting with the 'Blue Plan' that centred on a two axis advance to Kabul to the 'Pink Plan' of 1931 that was less ambitious and saw only a limited advance till Jalalabad and Kandahar. This was replaced by the Plan of Operations 1938 which envisaged only an active defence of the North West Frontier. Even in this plan, only one anti-aircraft battery was considered adequate and no new additions proposed. The AA battery formed part of the 'General Reserve'.[9]

The major change in anti-aircraft artillery in India came on the eve of the Second World War when the modernisation committee, under the chairmanship of Major General Auchinleck, recommended modernisation of the Army in India and a wider use of mechanisation. The report of the committee was used by the Chatfield Committee to firm up its recommendations and while doing so, the possible overseas employment of the India Armed Forces was also considered. Though the existing structure of military commands and districts in India was recommended to be retained, the Army in India was grouped under five categories, viz.;

Frontier Defence Troops,
Internal Security Troops,

Coast Defence Troops,
The General Reserve, and
The External Defence troops.

It was at this stage that first major changes were recommended for the anti-aircraft artillery, the first of having three anti-aircraft batteries as part of the 'General Reserve'. In a departure from past practices, these batteries were recommended to be of Indian Artillery and not the royal artillery as had been the norm till date.[10]

In a related development, a decision was taken by the Commander-in-Chief, India, in March 1938 to carry forward the process of Indianisation of the Army and enhance the number of Indian artillery units. Laying out the steps for the same, it was decided that having raised the first Indian field brigade, the process would start with the induction of Indian officers in the No. 1 Indian Mountain Artillery Brigade and it was to be followed by the replacement of a second British field brigade by an Indian field brigade. Thereafter, a British Heavy Battery (Coast Artillery) and the British Anti-Aircraft Battery were to be replaced by Indian batteries.[11]

When World War II broke out in 1939, the Indian Artillery, however, had no Anti-Aircraft units; notwithstanding the recommendations of having three AA batteries as part of the General Reserve and the replacement of the British AA battery by an Indian battery. The 8th AA Battery at Peshawar remained the only AA unit with the Army in India.

On the North West Frontier, there were fast-paced developments. Afghanistan had always been wary of Russian designs and had sought peace with its neighbour in the North. A trade pact was signed by the two sides in early 1940 but the success of an early German offensive, and the German-Soviet nonaggression pact made the danger of a Soviet invasion of Afghanistan clear and present. With the figures of 1,700 Soviet aircraft available and ready to carry out an offensive against Afghanistan doing the rounds, the Soviet air threat could no longer be ignored by both Britain and Afghanistan and the latter was keen on assurance of military aid from Great Britain. While the British Foreign Office reiterated in a memorandum on 3 April 1940 that Afghanistan was an integral part of British Imperial Defence and 'support must be given . . . by all means in our power', there were no forces available in India that could be stationed in Afghanistan though plans were prepared for an eventuality that the Army in India would have to undertake an advance into Afghanistan.[12] 'Plan A', prepared as per the directive issued by the Commander-in-Chief, India, on 21 May 1940 had the following military objectives:[13]

(a) To resist a Soviet attack by bombing Soviet aerodromes, installations, communications and troops beyond the Hindu Kush range, from bases in India.
(b) To support the Afghan government in the Kabul area with land and air forces via the Khyber pass.
(c) To forestall Soviet forces on the Helmand river and secure the Kandahar area with land and air forces via the Bolan pass.

With the advance planned along two lines, i.e. the Northern and Southern Lines with a General Reserve to be maintained in India, the requirement of anti-aircraft regiments was now assessed to be of three AA Regiments, of which one regiment each was proposed to be employed along the Northern and Southern Lines, with the third AA Regiment being retained at the Khyber pass as part of the reserve.

As the increased threat from the Soviet Air Force had been acknowledged, a need for static anti-aircraft defence of vital installations and military bases was also felt but the plan fell short of defining the exact numbers, only laying down the recommendation that the guns for static defence were to be provided by His Majesty's Government and that India was obligated only to provide the bulk of manpower for the static anti-aircraft defence.

The requirement of anti-aircraft regiments was revised in the follow-up plan in which four more regiments were recommended to be raised—one heavy mobile anti-aircraft and three light mobile anti-aircraft regiments—to serve with the expeditionary force. The guns for static defence were also finalized with the total requirement pegged at 234 static heavy and 153 static light anti-aircraft guns.[14]

THE PRELUDE TO WAR

The focus of the Army in India had remained largely on the North-West Frontier but it would be the developments in the Far East that were to have a far greater impact and would suck in the Army in the coming conflict.

The threat in the East was from Japan and for years not much was done by Great Britain to counter the growing military power of the Imperial Japanese Navy and Army.

After the end of the First World War, the Committee of Imperial Defence in 1921 examined the British defence policy in the Far East and recommended Singapore to be developed as a base to best serve the British interest and control the sea communications in South Pacific and the Indian Ocean. The new naval base on Singapore Island was proposed to be developed on the northern side in the Johore Strait with an airfield at Selatar. Work on the base did not start for years and even as Japanese forces landed in Shanghai in 1932, no action was taken to commence the construction of the naval base at Singapore. Only in 1933 was the authorization to start work given by the British government. With the first stage of defences at Singapore to be ready only in 1936 or 1937, it was already too late though it was proposed to develop two more airfields on the island and increase the number of gun emplacements. The plan was for Singapore to hold on for 90 days until the arrival of the fleet for its defence. In Malaya, three more airfields were authorized to be built but as the Air Force did not consult the Army while deciding on the sites, they were at places that were virtually impossible to defend. Though the focus had remained on Singapore, GOC Malaya was now authorized to begin constructing defences in Johore.

In 1938, with Italy aligning with Germany, not only did the threat to the communications with the Far East become real but there was a need for Britain to maintain a fleet at least equal to

Germany that would have to remain in its home waters. The need of sending a strong fleet to Singapore remained.

On 7 July 1937, war broke out between China and Japan. In October 1938, Japanese troops landed at Bias Bay, cutting off Hong Kong from the mainland. Early in 1939, it was decided by Great Britain to send a fleet to the Far East but because of threat from Germany and Italy, only two capital ships were sent. In August 1939, following the German invasion of Czechoslovakia, it was decided to reinforce Singapore with land and air forces. The 12th Indian Infantry Brigade Group from the 4th Indian Division was first to go to Singapore. The defences at Singapore at that time consisted of seven infantry battalions and five regiments of artillery deployed in two fire commands. For anti-aircraft defence, there were three regiments, including two of the Hong Kong-Singapore Royal Artillery.

Meanwhile, the coastal defences at Hong Kong had been modernised, though all the garrison had was two infantry battalions. For anti-aircraft defence, it had the 5th AA Regiment (previously designated as a brigade), with three AA batteries.

Burma was part of the Far Eastern Command and had a very small defensive force, called the Burma Auxiliary Force, made up of locals with both infantry and artillery component. No threat was envisaged from the east at this stage and all that Burma had for anti-aircraft defence was one AA regiment, but without any guns.

India had a large standing army but it was woefully antiquated. The modernisation process had just started in 1939, in line with the recommendations of the Chatfield Committee report and was yet to be completed. In 1940, India had sent the 4th Infantry Division comprising two infantry brigades to Egypt with the third brigade in Singapore. After Dunkirk, India was in the process of raising one armoured and five infantry divisions but there was a problem in providing artillery regiments for the new formations as an adequate number of units had not been raised as yet. There was just one unit—the 'A' Field Regiment with four 4-gun horsed batteries located at Bangalore (now known as Bengaluru)—one each of Punjabi Mussalmans, Rajputs, Ranghars and Madrassis.

It was with this as the backdrop that the first Indian Anti-Aircraft establishment came to be raised in 1940.

NOTES

1. N.W. Routledge, *History of the Royal Regiment of Artillery: Anti-Aircraft Artillery*, Brasseys, London, 1998, pp. 39-40.
2. Ibid., p. 59.
3. Bisheshwar Prasad, *Expansion of the Armed Forces and Defence Organisation, 1939-45*, History Division, Ministry of Defence, Government of India, New Delhi, 2012, pp. 8-9.
4. Indian Army List for July 1940, Defence Department, Government of India, Published by the Manager, Government of India Press, Delhi, 1941, p. 456.
5. Prasad, op. cit., p. 436
6. Ibid., pp. 10-11.
7. Routledge, op. cit., p. 59.
8. Prasad, op. cit., pp. 9-11.
9. Ibid.
10. Ibid., p. 24.
11. Brigadier Y.B. Gulati (Retd) and Maj Gen D.K. Palit, VC (Retd), eds., *History of the Regiment of Artillery—Indian Army*, Palit & Dutt Publishers, Dehradun, 1971, p. 46.
12. Sneh Mahajan, *Foreign Policy of Colonial India: 1900-1947* (Routledge Studies in South Asian History), Routledge, 2018, Kindle Edition.
13. Prasad, op. cit., pp. 439-41.
14. Ibid., p. 443.

CHAPTER 1

A Beginning is Made

The First Raisings

When the decision to raise the first Indian anti-aircraft regiment was taken in 1938, one of the first challenges was to decide on the class composition of the regiments and batteries. All artillery personnel, of Mountain Artillery and the Survey Sections, were recruited almost exclusively from Punjab with the class composition in the Mountain Artillery at the outbreak of the war being 50 per cent Punjabi Mussalmans, 38 per cent Sikhs, 8 per cent Ahirs and 4 per cent Punjabi Brahmins. The lone Indian Field Regiment was more broad-based in its composition as batteries were made up of Punjabi Mussalmans, Ranghars, Rajputs and Madrassis.

Overall, Indian artillery, as a whole, consisted of at least 88 per cent Punjabi Mussalmans and Sikhs, with an appreciable number of Ahirs, and a very small number of Punjabi Brahmins, Ranghars, Rajputs and Madrassis. This was the foundation available on which the Indian Artillery was to be expanded and there was an apprehension that any large expansion in classes other than Punjabi Mussalmans and Sikhs would lead to difficulties. This apprehension was misplaced but it would take time for the planners at the General Staff to realise the same. In any case, the problem lay elsewhere.

CHALLENGES

As all artillery units were dependant on the same recruitment pool, Punjab had become over-recruited with the intake beginning to fall off towards the end of 1940. It was now obvious that the traditional recruitment base for the artillery would not be able to cater to all the recruitment needs of the AA artillery. The need was thus to identify and tap a new base for recruitment, from classes other than the traditional 'martial races' or what was at times referred to as the 'pre-war classes'. One option was to tap the South Indian Classes (SIC), or the Madrassis, as they were called. The advantage with this approach was that the Indian artillery already had a field battery of Madrassis and was familiar with the men. There was, however, an associated challenge with this approach; that while they (the other classes) could provide a plentiful supply of recruits, there was no peace nucleus to build on. The main problem would be of providing the required number of Madrassi Indian Officers (VCO) and NCOs for the new regiments.

After weighing the available options, it was decided that the anti-aircraft artillery would expand to a limited extent from Punjabi masses but the main expansion will by introduction of Madrassi anti-aircraft regiments with a British cadre, to be found partly by milking units sent from home (India) but mainly from the NCOs sent from the UK specially for the purpose.

Another concern, reflected in some of the official correspondence on the subject, was the apprehension that too high a proportion of Madrassi might lead to 'letting the Army down'. This was more due to the misplaced concepts of the so-called 'martial races' than anything else and it is a different matter that the Madrassi units performed creditably during the war and never let the Army down.

Before the raisings could commence, there was a need for a core of officers, VCOs and NCOs who were trained on anti-aircraft techniques and could help in the raising of the new anti-aircraft regiments. Towards this end, a Coastal Defence & Anti-Aircraft School was set up at Karachi in June 1940 that turned out the initial group of trained personnel. This was followed by the raising of No. 1 Technical Training Battery on 15 September 1940, using the personnel of the 8th AA Battery, RA. This battery included the core group of the 'R' Heavy Anti-Aircraft Regiment that was raised on 1 October 1940, at Bombay.[1] A beginning had been made but the challenges remained. The decision of using other classes for anti-aircraft artillery had resolved only one issue. Equally important was the provision of VCOs and NCOs for the new regiments.

In the initial raisings, the VCOs and NCOs were taken from the Mountain Artillery as it had the only readily available pool. This experiment soon ran into trouble owing to language difficulty, difference of outlook and lack of sympathy, and led to considerable unrest in various units. While Urdu was the preferred language in the artillery, the south Indian gunners were more conversant with English and preferred it over Urdu. The Punjabi VCOs and NCOs from Mountain Artillery found it difficult to communicate with the Madrassi gunners leading to avoidable problems. Cultural differences, including food habits, were another contributing factor in creating a chasm between the VCOs and NCOs, and the gunners.

With the outbreak of war with Japan, Mountain Artillery had to send some units to Burma while others were already in Malaya and the Middle-East. These units were unable to give any assistance to the expansion of the artillery in the manner they had been earlier. The provision of NCOs by the Mountain Artillery units was now putting a strain on them and beginning to affect their operational and administrative efficiency. It was becoming obvious that Mountain Artillery could no longer support the expansion of anti-tank and anti-aircraft artillery. On the other hand, the need for more anti-aircraft units became urgent as the danger to India was growing from both East and West.

At one stage, it was contemplated that suitable NCOs could be taken from other arms but as they (the other arms) themselves were in a process of expansion, no NCOs were available for transfer to artillery. As a result, anti-aircraft artillery went in for 'mixed' units from 1942 onwards—made up of British NCOs and Madrassi gunners. This was followed by a policy of direct enlistment of Madrassi VCOs and NCOs. This policy also meant that the British NCOs could be relieved of their duties in Indian regiments and batteries, and be employed elsewhere. Over a period of time, with training and experience, the VCOs and

NCOs in the regiments came up from the Madrassi cadre itself, which was preferred over 'importing outsiders'.

The provision of officers for the anti-aircraft regiments proved equally difficult as the requirement was of officers not only proficient in anti-aircraft techniques but also with experience of Indian troops and knowledge of Urdu. The problem was accentuated in the early years as a large number of RA officers, including from Mountain Artillery and with good experience of Indian troops, were sent to England just before the start of the war to act as instructors for the militia. The officers on leave from India, at the outbreak of war, were retained in the UK and a number were sent from India during the first six months of the war. There were, however, a few officers in other types of units who had previous service with anti-aircraft units and such officers were combed out from their units and transferred to anti-aircraft artillery. However, the general shortage of artillery officers in India continued for almost two years.

Manpower was not the only concern for there was shortage of equipment and transport. There were no guns to begin with; 8th AA Battery, RA, had given four of its guns to the AA battery of the Burma Auxiliary Force and had a section at Aden. That left hardly any AA guns that could be spared for the new recruits to be trained on till the arrival of guns from UK. Lack of transport was another concern as all existing artillery units were on animal basis with no mechanical transport. Training on and using mechanical vehicles would have been a novel experience for almost all recruits. The general low standard of education in the classes from which recruitment was done was yet another problem though it was felt that the South Indian classes would be better educated as compared to the 'traditional' classes, and hence, it would not be a major problem for the anti-aircraft artillery units.

TRAINING CENTRES

In 1939, there was only one training centre for Indian artillery—the Mountain Artillery Training Centre (MATC) at Ambala that was responsible for training all recruits for Mountain Artillery. This was an old established training centre having recruitment connections in the Punjab from whence all recruits came except some Ahirs from the vicinity of Delhi. The centre contained only two training batteries, one each for training gunners and mule drivers. During the first two years of the war, it also trained recruits for coastal artillery, anti-tank artillery, heavy and light anti-aircraft artillery and survey units.

The other training centre, at Mathura, was for the Royal Artillery which had the additional responsibility of training Indian drivers for British field and medium regiments, nearly all of which were horsed. The Royal Artillery Training Centre (RATC) had also been given the task of training recruits for the first Indian field regiment ('A' Field Regiment, later designated 1 Field Regiment). At the outbreak of the war, its main connection thus lay with the Royal Artillery rather than with the Indian Army with the limited role of training only a handful of Indian personnel as horse drivers. The facilities for training gunners and other personnel were naturally limited as the peace wastage for one field regiment was small. The Field Artillery Training Centre (FATC), formed a year after the start of the war from the old RATC, originally trained recruits of field artillery only, but later took over from MATC the task of training recruits for anti-tank artillery as well.[2]

With only one AA battery of Royal Artillery located in India, the RATC was responsible for imparting anti-aircraft training to British personnel as well. During 1940-1, it trained a considerable number of recruits for anti-aircraft artillery until the Anti-Aircraft Training Centre (AATC), formed in January 1941, was able to stand on its own.

The system of initial training of recruits being followed was that the recruits received their initial training at the artillery training centre for an average period of six months and were then drafted to new units for which NCOs were pooled in from existing units. The number of recruits at the training centre as per this system was about half the number of gunners in the establishment of all the units that are to be raised during the year.

Due to the large expansion planned for the anti-aircraft artillery, the existing training centres need to be very large—or new centres were required to be formed. Very large training centres ran the risk of being inefficient while new training centres demanded a greater overhead in personnel, both officers and NCOs. Though both the MATC and FATC were expanded early on in the war, it was decided that a separate training centre would be formed for the anti-aircraft artillery. As the first step towards having a separate establishment for anti-aircraft artillery, an Anti-Aircraft and Coastal Defence Wing was established in June 1940 by the School of Artillery, Kakul, near Rawalpindi for training of Indian officers, VCOs and NCOs in AA techniques and for refresher training of British All Ranks transferred to India from the AA Command in Britain. This was followed by the raising of the No. 1 AA Technical Training Battery at Gun Carriage Basin, Colaba, Bombay. The Training Battery moved to Karachi in December 1940, and was renamed as the first AATC in 1941, with the addition of another training battery. All raisings and training of recruits now became its responsibility though recruit training of select units of AA Artillery continued to be carried out at the FATC, Mathura.

THE FIRST RAISINGS

The first Indian anti-aircraft unit to be raised was the Anti-Aircraft Training Centre on 15 September 1940 at Karachi. This was formed by a nucleus drawn from the 8th AA Battery, RA, which had reported to Gun Carriage Basin, Colaba, Bombay, for the new raisings. Part of this core group of AA Battery moved to Karachi, to form the AA Training Centre, while the personnel left behind at Bombay were used to form the first regiment, i.e. the 'R' Indian Heavy AA Regiment on 1 October 1940. With this, the Indian Artillery finally had an AA regiment on its order of battle. The 'A' Battery (later designated as the 1st Indian HAA Battery) of 'R' Indian HAA Regiment was formed only on 1 March 1941 at Malir Cantt. The delay in raising of the regiment and forming of the first battery was due to the fact that the recruits underwent basic military training for two months under their British NCO instructors and then formed the batteries with the same NCOs, undergoing technical and operational training for three to four months after which the batteries were grouped together as regiments. The regiment may have been raised on 1 October 1941, the first battery was 'formed' only in March 1941.

While the regiment was being raised, the School of Artillery moved from Kakul to Deolali on 1 January 1941 and opened a cadet wing for British and Indian potential officers. This was necessitated due to the shortage of captains and subalterns for the new raisings. Later, as with other arms, the potential officers went to the Indian Military Academy at Dehradun and did a short course after which they were granted emergency commission.

The next anti-aircraft regiment to be raised was the 'U' Indian Light AA Regiment by Major Hawkes in January 1941, at Malir Cantonment near Karachi. The 'U' Indian Light AA Regiment had the 2nd, 3rd and 6th Indian Light AA Batteries as 1st, 4th and 5th Indian Light AA Batteries were retained as independent batteries, of which the 5th Indian LAA Battery was the first to be raised with Madrassi gunners. The batteries of 'U' Indian LAA Regiment were a mix of Punjabi Brahmins (2nd Indian LAA Battery) and Punjabi Mussalmans (3rd and 6th Indian LAA Batteries).

One of the first officers to be posted to the AATC, and then to 1 Indian HAA Regiment was Major Charles H.T. MacFetridge, a mountain gunner who had served in the NWFP. He had been selected due to his experience with Indian troops and knowledge of Urdu. He recollects that the recruits were 'extraordinarily raw' who had to be taught everything from scratch but eventually 'turned out to be very good soldiers'. He was posted to the 1st Indian HAA Battery and was involved with the initial training of the recruits when he received orders in April 1941, to proceed to Delhi with all his kit and report to the Royal Artillery Directorate. On arrival at Delhi, he was told that, rather than being briefed by the Royal Artillery Directorate, he would be interviewed by Major General Scoones, Director, Military Operation at the GHQ, India. General Scoones informed him that he was to proceed to Digboi and reconnoitre suitable gun sites for AA guns, to be deployed for the anti-aircraft defence of the oil refinery. This was well before there was any public admission of an impending war with Japan, and the raising of the 'R' Anti-Aircraft Regiment was still not complete with the training ongoing at Malir.[3]

MacFetridge, on arrival at Digboi, carried out the reconnaissance to select the gun sites but was hampered by lack of any administrative support from military unit(s) as he had been told to be discrete and carry out the reconnaissance in civil clothes so as not to cause any alarm amongst the civilians. Even otherwise, the gun sites were difficult to be selected owing to the terrain and thickly forested area all around. The first two sites selected turned out to be sites earmarked for construction of the bungalow for the manager of the oil refinery and the lawn tennis court for the local club. At last, two sites were selected and confirmation sent to the GHQ for the same. Orders were issued for the gunners to be sent to Digboi, followed by the AA guns. When the 1st Indian HAA Battery arrived in Digboi, the men had barely completed their training and all they had was five 18 pounder Mk I guns, of World War I vintage, to be used in anti-aircraft role. They were deployed at two sites—three guns at one and two at another.

The first battery of Indian anti-aircraft artillery was thus the first to be deployed for war though it did reveal the lack of preparedness to counter any real air threat. Not only were the men undertrained but even the guns were of 1920s vintage; though that was all that could be mustered at that stage.

This was followed by the raising of the 'S' Indian Heavy AA Regiment. In April 1941, at Drigh Road, Karachi, with the 7th and 8th Indian Heavy AA Batteries. Interestingly, the 5th and 6th Indian Heavy AA batteries were not part of 'S' Indian HAA Regiment and formed part of the 3rd Indian Heavy AA Regiment that was raised later in June 1941.[4]

The initial lot of officers posted to the Indian AA Artillery were preferably those who had not only served with Indian troops but also knew Urdu, as it was accepted that Urdu would be used as the 'official' language for communicating with the troops by the British officers. However, the shortage of such officers meant that the young officers posted to these units had no knowledge of Urdu, and had to undergo a course to learn the language. This may have been the official policy but in the initial raisings with Punjabi troops—Punjabi Mussalmans, Sikhs and Punjabi Brahmins—the common language used in the regiments was Punjabi, and not Urdu.

On 2 April 1941, the nomenclature of the anti-aircraft regiments underwent a change with numeral designation being used. Accordingly, the 'R' Indian HAA Regiment was now called '1st Indian HAA Regiment', the 'S' Indian HAA Regiment became '2nd Indian HAA regiment' and so on. Interestingly, there was a duplication in designation of heavy and light AA regiments; both using the same numerals for designating the regiments.[5]

During the year, three more regiments were raised at Karachi, viz., 3rd Indian HAA Regiment on 1 June, the 2nd Indian LAA Regiment on 1 October and the 4th Indian HAA Regiment on 1 November 1941, making it a total of six Indian AA regiments and four independent AA batteries on the order of battle at the time Japan declared war on 7 December 1941. The Indian AA artillery had no brigade headquarters at this time.

The first Indian AA unit to be sent overseas was the 1st Indian HAA Regiment that set sail for Malaya on 19 July 1941 from Karachi. In addition to its three HAA batteries, the regiment had the 1st and 5th Indian LAA Batteries affiliated with it. After disembarkation, it moved to Chak Abu camp near Nee Soon to set up camp under canvas but shifted to a large hutted camp that was found nearby. The regiment was soon busy taking over guns, vehicles, equipments and the anti-aircraft gun sites at the naval base, airfields, and administrative and logistics depots in Singapore that they had been assigned by the Malaya Command.[6] It was going to be a hectic time ahead for the regiment as it geared up to face the challenges ahead. But before the 1st Indian HAA Regiment was to be tested in battle, the anti-aircraft battery of Hong Kong Singapore Royal Artillery, manned by Indians, faced the Japanese offensive in Hong Kong. These were truly the first Indian AA gunners who went into battle.

NOTES

1. The commonly referred to date of raising is 14 September with No. 1 Technical Training Battery being formed on this date at Colaba. However, the date of raising of the first AA unit, as recorded in official documents, is 15 September with the place of raising recorded as Karachi, and not Colaba, Bombay.
   Statistical Review of Personnel, Army of India, vol. I, 1 July 1939 to 1 October 1941, compiled by Statistical Section, GHQ, Delhi, 1941.
2. Notings of Artillery Directorate on 'Expansion of Artillery 1939-1941' accessed at History Division, Ministry of Defence, New Delhi.
3. Recollections of Colonel Charles MacFetridge, Reference Number 21295, Oral History, Imperial War Museum, London.
4. *Statistical Review of Personnel*, op. cit.
5. Brigadier Y.B. Gulati (Retd.) and Maj Gen D.K. Palit VC (Retd.), eds., *History of the Regiment of Artillery—Indian Army,* Palit & Dutt Publication, Dehradun, 1971, p. 52.
6. File No. 601/2290/WD, War Diary of 1st Indian HAA Regiment maintained by History Division, Ministry of Defence, New Delhi.

CHAPTER 2

Hong Kong-Singapore Royal Artillery

The First Experience with Anti-Aircraft Artillery

Much before the first Indian AA regiment was raised in October 1940, Indian gunners were already serving with anti-aircraft units in the Far East and to trace the history of the first Indian anti-aircraft gunners, we need to look eastwards at the British garrisons of Hong Kong and Singapore.

In 1841, four companies of the Madras Army were sanctioned to be formed at Fort St George and sent to Hong Kong for the defence of the newly occupied territories. The first reference of these companies, called the 'China Gun Lascars', is available in *The Asiatic Journal and Monthly Register for British and Foreign India, China and Australasia* of 1842 which mentions the newly raised companies as follows:[1]

> Under instructions from the Right Honourable, the Governor in Council, four companies of gun lascars have been raised for service in China, each consisting of 1 subedar, 1 *jemadar*, 8 *havildars*, 115 *lascars* and 2 *bheasties*. The companies are to be lettered from A to D, and will be under the orders of the officer commanding the Madras Artillery in China for the general duties of the Corps. Officers in command of these companies are authorized to draw an allowance of Rs. 30 a month for stationer, *&c.*

The companies were commanded by Subedars, with no British or native officer on strength of the company. The men dressed like all other garrison gunners though their duties seem to have been purely orderly and fatigue. The men were initially recruited from (the then) Madras Presidency but the recruitment base was later expanded to include Punjabi Mussalmans (PM) from Punjab, and a Punjabi Mussalmans' battery was duly raised as a one-class company.

The class composition of the lascar companies underwent a major change in 1865 when men of all Asiatic nations were allowed to be enlisted. These included Portuguese half-castes and Jews and they were formed into mixed companies though the Punjabi Mussalman company was retained as such. The duties of the lascars were slowly expanded to include gun drill and musketry with the men showing signs of efficiency. Over a period of time, the fatigue duties were restricted and the focus shifted to gunnery duties.

In 1881, an additional company was sanctioned following the increase in armament in Hong Kong and at this stage, Colonel Hall, the then Officer Commanding Royal Artillery, Hong Kong, expressed a desire that the new company should be composed entirely of Sikhs enlisted in the Punjab, India. The companies were still called the 'China Gun Lascars'. This led to discontent amongst the Indian gunners, especially the newly recruited Sikhs who had been enlisted in India for the *Top Khana* (Artillery), but it was not till 1891 that the name was changed from China Gun Lascars to 'Asiatic Artillery'. By now, there were 'Asiatic Artillery Companies' in Hong Kong, Singapore, Ceylon, and Mauritius.[2]

To bring about greater control and efficiency, the 'Asiatic Artillery' underwent a reorganisation in 1898 and two separate battalions were formed: 'The Hong Kong and Singapore Battalion, Royal Artillery', headquartered in Hong Kong, and the 'Ceylon Battalion, Royal Artillery', headquartered in Ceylon.[3] The next year, the Royal Artillery was split into two distinct branches with Royal Garrison Artillery coming into being. With this, the two battalions were now named the 'Hong Kong Singapore Battalion, Royal Garrison Artillery' and this arrangement lasted till 1924 when it was decided to again merge the two, i.e. the garrison and field artillery, into a single corps.

In 1934, the units underwent another change and were to be called 'Hong Kong-Singapore Royal Artillery' (HKSRA). It now consisted of one mountain and three heavy batteries at Hong Kong, and one heavy battery at Singapore. With the need for anti-aircraft defence of Singapore and Hong Kong, the 5th AA

Brigade (later designated Regiment) was formed in 1934 with a HQ and the 7th AA Battery of Royal Artillery.[4] Two anti-aircraft batteries of HKSRA (Nos. 7th and 9th) were raised with Punjabi Mussalmans and Sikhs recruited from Punjab in India, who joined the brigade after training.

These were the first anti-aircraft batteries manned by Indian gunners.

By September 1939, HKSRA had grown into two mountain batteries, two medium batteries, and one heavy battery at Hong Kong, six anti-aircraft batteries at Singapore and one heavy battery at Penang. The AA batteries were organised into 1st and 2nd AA Regiments, HKSRA. In 1940, the HKSRA was rocked by a mutiny when all personnel were ordered to wear steel helmets instead of *pugris.* The Sikh sepoys of the 20th Heavy Battery of the 12th HKSRA, along with the 2nd Battalion of the 14th Punjab Regiment, refused to wear the helmets. The mutiny spread with 83 Sikh sepoys of the 20th Heavy Battery displaying collective insubordination on 19 December and the Sikhs of Hong Kong Rifles refusing to handle the crates containing the steel helmets. They were joined by Sikhs of the 5th AA Battery, refusing food. The mutiny was quelled only by a collective court martial of 83 Sikh gunners in January 1941.[5]

The mutiny and insubordination notwithstanding, the HKSRA acquitted themselves with honour once hostilities were joined with the Japanese invasion.

BATTLE OF HONG KONG

Hong Kong, the first to face the Japanese onslaught, had only one anti-aircraft battery, the 17thAA Battery of HKSRA which formed part of 5th AA Regiment, Royal Artillery. In addition, the regiment had the 7th and 18th AA Batteries of Royal Artillery and the 5th AA Battery formed from the Hong Kong Volunteer Defence Corps.[6]

The Japanese attacked Hong Kong on 8 December 1941 with two divisions supported by 80 aircraft and a large component of corps artillery. During the battle, Japan lost a total of eight aircraft; one of which was brought down by the 17th AA Battery, HKSRA, during the Battle of Sai Wan. The battery suffered losses at Brickhill and Bluff Head when sections where knocked off by concentrated Japanese air attacks.

Faced with overwhelming odds, the British garrison in Hong Kong surrendered on 26 December. After the fall of Hong Kong, *sipahis* from five Indian units were captured, the largest proportion of whom were classified as the PM (Punjabi Mussalman) from HKSRA, the 2nd/14th Punjab and 5th/7th Rajputs. The HKSRA alone had 1,500 *sipahis* taken as prisoners of war. As the Punjabi Mussalmans were considered, more likely than other Indian *sipahis,* to join the Indian National Army (INA), they were separated from the others and were kept in the Gun Club Hill Barracks. This group was later joined by more than 800 Punjabi Mussalman *sipahis* in May 1942. The Punjabi Mussalmans were given better rations and the Japanese even encouraged them to mix around with each other and have 'free intercourse'. The efforts of the Japanese and Indian National Army to try and recruit them did not yield the desired results and in the end, only 100 of these agreed to bear arms under the Japanese, and were thereafter sent to Guangzhou in mainland China as guard pickets.[7]

The gunners of anti-aircraft regiments of the HKSRA were sought after more by the Japanese for use on air defence duties and Iwakuro Kikan, the local Japanese commander at Singapore, asked the INA commander, Mohan Singh, to try and recruit the anti-aircraft gunners for the Japanese units. These men were not placed in INA units but were placed directly under Japanese officers and were given Japanese Army uniforms. While some of them were sent to Celebes for garrison duties in the Dutch East Indies, the luckier of the lot were used for air defence duties only.

Some of the anti-aircraft gunners of the HKSRA who had joined the Indian National Army were deployed on Air Defence duties, manning anti-aircraft guns at and near Singapore after the Japanese takeover.[8]

MALAYA

The situation in Malaya and Singapore was similar, of unfinished defences and inadequate forces. Long neglected, the measures undertaken to beef up the defences of Malaya and Singapore proved to be too late and too little. The anti-aircraft defences in Malaya included the 1st and 2nd AA Regiments of HKSRA and the HAA batteries were initially deployed at the forward airfields. Alor Star aerodrome had 4 × 3" guns of the 9th HAA Battery with four more at Kota Bharu. Sungei Patani had eight static 3.7-inch guns of 13th HAA Battery and two guns were deployed at Gong Kedah. The aerodrome at Tengah was the most heavily defended with the 10th HAA Battery and the 20th HAA Battery deployed for its defence; as was the 4th HAA Battery of the 1st Indian HAA Regiment.

An ad hoc Light AA Regiment, called the 3rd Light AA Regiment, HKSRA, was made up using the available light AA batteries in Malaya, elements of which were with the Indian III Corps, split up into troops and sections for anti-aircraft defence of the brigades.[9] The detailed deployment of the anti-aircraft regiments of HKSRA was as follows:

1st HAA Regiment, HKSRA
- 6th HAA Battery at Johore Bahru and Sungei Buloh
- 9th HAA Battery at Johore State and Naval Base
- 10th HAA Battery at Tengah and Mount Vernon

2nd HAA Regiment, HKSRA
- 12th HAA Battery at Nee Soon and Holland
- 13th HAA Battery at Geylang and Club
- 20th HAA Battery at Tengah

3rd Light AA Regiment, HKSRA
- III Indian Corps

Though the AA defences were later augmented by the arrival of British AA regiments, they all proved inadequate in the end.

The Japanese invaded Malaya on 8 December 1941 with the troops landing at Kota Bharu and Singapore had its first air raid in the same day as 17 Japanese planes bombed various bases and installations on the island. Havildar Udham Singh of 18th LAA Battery, HKSRA, was deployed at Stanley Battery site when his detachment as ordered to engage a landward target. While firing, one of the projectiles broke in front of the muzzle and a fragment fell near the ammunition stored at the gun site. As the ammunition caught fire, Udham Singh, at great risk to his personal safety, managed to put out the fire and save not only the AA gun but also his detachment.[10]

The gunners of HKSRA acquitted themselves with honour during the battle on the mainland but it was not enough to hold back the Japanese, who overran the defences on the mainland in little less than two months and attacked Singapore on 8 February. The invasion was accompanied by intense air raids, with relentless targeting of defences and installations by the Japanese planes and the anti-aircraft guns came in for special attention. On more than one occasions, the HKSRA gunners rose to the challenge, taking on more responsibilities than their call of duty. One such gunner was Naik Mohd Amir who was a GPOA of a gun position at the Golf Club in Singapore. When Gun no. 1 was injured in the constant shelling and bombing, Mohd Amir took on the responsibility of the gun commander and maintained the gun in action.[11]

Such acts of bravery were not an exception as almost all personnel displayed exemplary gallantry in facing the challenge but as in the case of Hong Kong and mainland Malaya, it was all futile in the end.

After a brief resistance, the garrison surrendered on 15 February 1942.

During the battle, anti-aircraft batteries of HKSRA are credited with having shot down 51 Japanese planes with 30 probable hits. The anti-aircraft gunners of HKSRA won a Military Cross and two Military Medals during the Malaya Campaign. These included Jemadar Amar Singh of 2nd Light AA Regiment, HKSRA, who was awarded the Military Cross. He was in command of a 40 mm. LAA gun at the Police Barracks, Mount Pleasant, when his detachment shot down a Japanese plane in face of repeated and persistent Japanese air raids.[12] The other awardees were

Havildar Udham Singh, MM of 18th Light AA Battery, HKSRA, and Naik Mohd Amir, MM of 1st AA Regiment, HKSRA, who were awarded the Military Medal. The honours came with a heavy price for the anti-aircraft batteries of HKSRA which lost over 160 gunners during the battle with over 100 dying later in the prison camps.[13]

NOTES

1. *The Asiatic Journal and Monthly Register for British and Foreign India, China and Australasia,* vol. 38, 1 January 1842, pp. 48-9.
2. Barry Renfrew, *Forgotten Regiments: Regular and Volunteer Units of the British Far East,* Terrier Press, Amersham, 2009.
3. Ibid.
4. N.N. Routledge, *History of the Royal Regiment of Artillery: Anti Aircraft Artillery,* Brasseys, London, 1998, p. 59.
5. Kaushik Roy, *Sepoys Against the Rising Sun: The Indian Army in Far East and South-East Asia, 1941-45,* Brill, London, 2016, pp. 56-7.
6. Routledge, op. cit., pp. 2015-16.
7. Gajendra Singh, *The Testimonies of Indian Soldiers and the Two World Wars: Between Self and Sepoy,* Bloomsbury, 2014 (Kindle Edition).
8. At least two gunners of HKSRA are mentioned in the records as having joined the AA battery of INA that was deployed in Burma. One of them served as a Lieutenant while the other was made a Havildar in the Anti-Aircraft Regiment. Both died in Burma in 1945.
9. Routledge, op. cit., pp. 218-20.
10. *London Gazette* dated 4 April 1946 (Reference WO 373/47/365).
11. *London Gazette* dated 1 August 1946 (Reference WO 373/47/287).
12. *London Gazette* dated 13 December 1945 (Reference WO 373/47/201).
13. Roll of Honour of Kranji War Memorial, Singapore, maintained by Commonwealth War Graves Commission.

CHAPTER 3

Baptism

The Campaign in Malaya, 1941-1942

In 1932, the British government, noting the aggressive actions of Japan, decided to formulate a policy for defence of its interests in the Far East. That policy was based on the strategic thinking of the Royal Navy and was centred on Singapore. A battle fleet would be sent there to forestall any threat of invasion and Singapore was to be made into a fortress with a garrison, heavy coastal guns up to 15-inch and AA defence. The necessary programme of work was to be completed by 1937. At that time, the Japanese air threat took second place in estimation to the German one. Japanese military successes in China lent a spur to the defence programme: it was decided to allocate 3.7 and 4.5 inch AA guns as soon as these could be produced. The coast defence searchlights were to be supplemented by an AA layout and it was also planned to equip key points in the naval base with light AA guns.[1]

Up to this point, no serious thoughts seems to have been given to the defence of the mainland of Malaya but two factors now appeared to alter this: first, the possibility of Japan's invading that country from Indo-China by way of Siam, and second, a decision to rely on immediate interdiction by an RAF striking force before the fleet's arrival. The RAF, therefore, now planned new airfields along the eastern coast and in the north of Malaya before their means of defence had been fully considered. From 1937 to 1939, Maj Gen W.G.S. Dobbie, GOC Malaya, faced with this new situation, carried out studies to determine how to hold key routes through the central jungles, as well as the airfields, up to the Siamese border, a distance of some 400 miles from Singapore Island. He also stated that a seaborne attack would be feasible during the northeast monsoon, that is, from October to March.

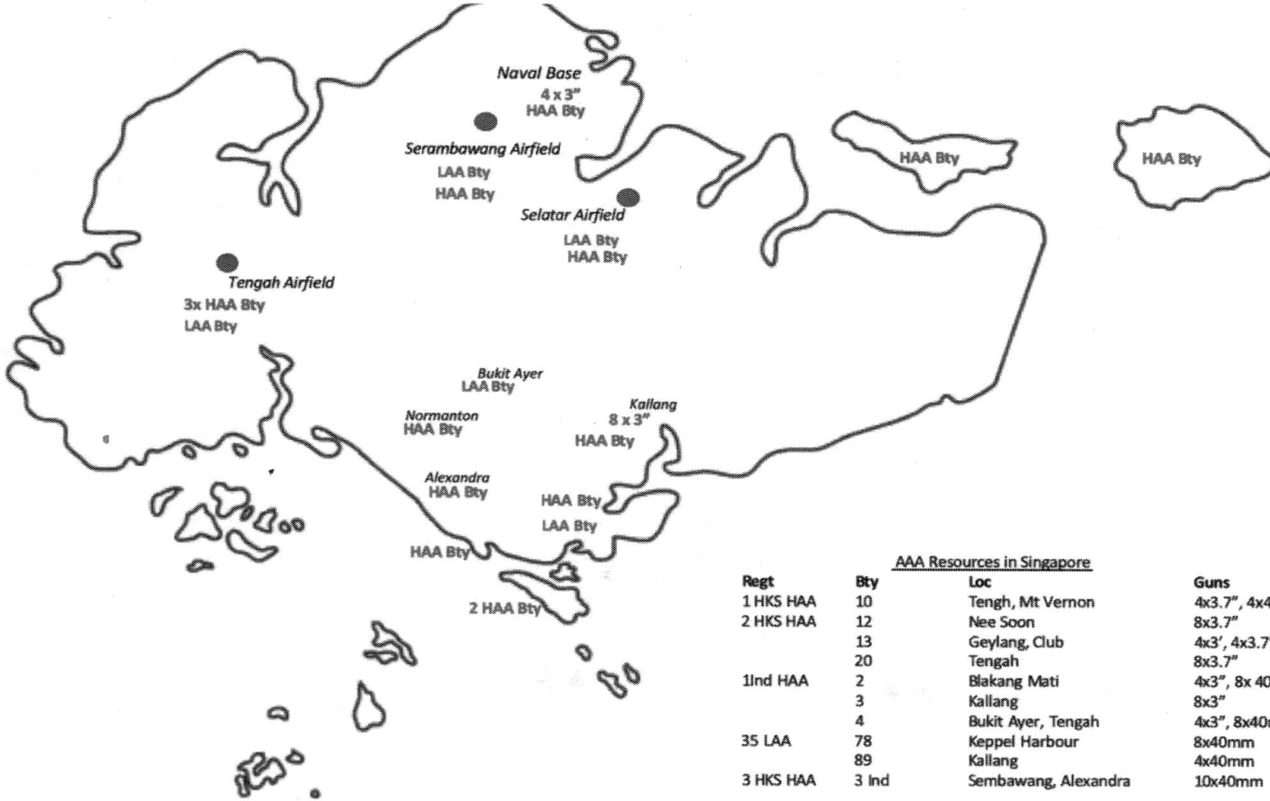

AAA Resources in Singapore

Regt	Bty	Loc	Guns
1 HKS HAA	10	Tengh, Mt Vernon	4x3.7", 4x4.5"
2 HKS HAA	12	Nee Soon	8x3.7"
	13	Geylang, Club	4x3', 4x3.7"
	20	Tengah	8x3.7"
1Ind HAA	2	Blakang Mati	4x3", 8x 40mm
	3	Kallang	8x3"
	4	Bukit Ayer, Tengah	4x3", 8x40mm
35 LAA	78	Keppel Harbour	8x40mm
	89	Kallang	4x40mm
3 HKS HAA	3 Ind	Sembawang, Alexandra	10x40mm

Map 1: AA Defences in Singapore. *Source:* Author's collection.

A plan was also prepared by, among others, Lt Gen J.V. Bond, Dobbie's successor in 1940, for entering no place else than Siam to pre-empt the use of ports and railways useful for an invader but, because of Siam's neutrality, this could only be contemplated if invasion was seen as imminent.[2] Lt Gen A.E. Percival, who succeeded Bond in May 1941 as GOC Malaya estimated that five divisions and more AA guns would be required to implement these plans.[3] Troops could only be found from India and, though one corps of two divisions was sent ultimately, but in 1939-41, AA guns were simply not available as reinforcements; and the demands for home defence and the Mediterranean had priority.

Less attention was paid to Hong Kong in defence planning as it was not itself a fleet base; the island and the single airfield at Kai Tak, Kowloon, were regarded as a form of outpost in Far Eastern strategy, to be held as long as possible. Moreover, the Foreign Office, even as late as 1941, seemed to discount the extent of Japanese ambition in the Far East and its intelligence reports were certainly at fault. However, Hong Kong was to be fortified and given a garrison with emphasis on coast defence and a very limited air defence capability. Much of the artillery was manned by the HKSRA. There was, in addition, a Volunteer Defence Corps.[4] The British government regarded Hong Kong as expendable and did not consider it a prime Japanese objective.

Events in Europe, in particular the fall of France in 1940, badly disrupted the British strategy for the defence of Malaya. The Italians now had to be contained in the Mediterranean without French help, so it was no longer possible to send a battle fleet to Singapore at short notice. On top of this, the Japanese profited by the French collapse to enter Indo-China, demand the use of airfields and the passage of the troops through Siam where there was a strong pro-Japanese faction. At the same time, they demanded supplies of war material from the Netherlands East Indies. Protracted negotiations to limit Japanese expansion, led by America, failed in November 1941; military leaders in Tokyo had already set a secret deadline for full-scale attacks the next month.

From their meticulously planned and concerted attacks, on Hong Kong, Singapore, the Netherlands East Indies and the

Philippines, together with Pearl Harbour, designed to sweep the British and the Dutch out of the Far East while neutralising the United States in the Pacific, the Japanese assembled 11 divisions, well supported by artillery and some armour and backed by powerful fleets. To cover operations on land, about 700 first-line aircraft were available for direct support of the Imperial Army; there were also about 500 carrier borne and land-based aircraft of the Imperial Navy which would be employed in the campaign. One infantry division was assigned, with air support, for the capture of Hong Kong, four divisions, together with an air division, were allotted for the invasion of Malaya and capture of Singapore and the remainder were destined for the East Indies, the Philippines and Siam. The operation against Pearl Harbour was entirely in the hands of the Navy. In the five or six years of training and preparations preceding these invasions, the (Army) Air Force had modelled itself closely on the Luftwaffe in its tactics. The aircraft in use were modern and had good performance; attention had been paid to giving bombers long operational ranges. Tactical handling included the close support of frontline troops with low level and medium level dive bombing as well as the level bombing of targets. Japanese pilots were generally efficient though relying heavily on those leading formations for their navigation. Attacks would be pressed home with fanatical energy regardless of the heaviest fire and this put a premium on the defence's destroying aircraft rather than relying on deterrence. Five air divisions existed for the army's support by 1941 but some were earmarked for home defence and for service in China. The 3rd Air division, assigned for Malaya, was specially augmented by 100 heavy bombers comparable to Ju-88s.

Against this air threat, the RAF was in a poor position. Plans demanded a force of 336 fighter, bomber and torpedo carrying aircraft for Malaya, a figure later raised to 582, but in 1941, the Air Ministry had sent only 158 and many of these were obsolescent. The torpedo carriers, for example, were the Wildebeestes, capable of about 100 mph., and many of the fighters were Brewster Buffaloes, inferior in every way to the Japanese 'O' type. Hong Kong had only five aircraft at Kai Tak; three torpedo carrying

Wildebeestes and two Walrus amphibians; there was no fighter at all.[5]

Such was the general state of defences when hostilities abruptly started on 7 December 1941.

HONG KONG

The garrison and defence forces of Hong Kong, in the late 1941, consisted of six infantry battalions, the 1st Hong Kong Singapore Regiment, RA (1st HKSRA), the 5th AA Regiment, RA, and the 8th and 12th Coast Regiments, RA. The AA regiments was made up of four batteries: the 7th and 18th with mainly British gunners, the 17th HKSRA Battery and the 5th AA Battery from Hong Kong Volunteer Defence Corps. It had a mix of 3-inch, 3.7-inch and 40-mm. AA guns. The regiment had its own Gun Operations Room (GOR) and the Commanding Officer was the Anti-Aircraft Defence Commander (AADC). There was only a small presence of RAF.[6]

The AA positions were sited to defend the harbour area, docks and coast defence guns as also the central pass in the interior of the island, Wong Nei Chong Gap. New sites were prepared for anti-aircraft guns at Mount Davis, Kellet Bay, Stanley and Chung Hom Kok. All that Hong Kong had for fixed anti-aircraft defence were sixteen anti-aircraft guns eight—3-inch semi-mobile, four 3.7-inch guns and four 4.5-inch AA guns, with two guns deployed at all sites, except Sai Wan where four 3-inch AA guns were deployed. With these two-gun sites, the area covered by AA defences was inconsequential and it hardly provided any effective air defence to the garrison. The light AA defence was provided by the six 40-mm. guns, along with a number of light machine guns. In all, a token force, more than anything else.

The first air raid came on 8 December about 8:00 a.m. that destroyed the motley group of RAF aircraft at the Kai Tak Airport.[7] This was accompanied by the assault across the Sham Chun River by six battalions of the Japanese 38th Division. No match to the invading Japanese, the last Commonwealth troops on the mainland had retreated to Hong Kong Island by 13 December.

The island garrison did not hold for long and the battle was all but over by the afternoon of 25 December 1941, when the Governor of Hong Kong surrendered in person at the Japanese headquarters. This was the first occasion on which a British Crown Colony had surrendered to an invading force. The garrison had held out for 17 days.

The battle was doomed from the start for the Allies as they faced a much superior and numerically larger force. The complete absence of an Air Force, combined with the miniscule AA defences, meant that the Allies had no chance of even putting up a decent fight. And this was only the portend of things to come, for the Japanese soon turned their attention towards Malaya and the prize catch of Singapore.

MALAYA

The general defence plan was based on an appreciation written by the Chiefs of Staff in July 1940 which laid down that defence was to rely, in the absence of a fleet, primarily on air power.[8] Based on this paper and the subsequent tactical appreciation, it was recommended that a minimum strength of 582 aircraft be maintained to meet the defence commitments in the Far East. To house them, there was a need for the construction of adequate number of bases, of which 16 new ones were to be built in Malaya. Of the bases in existence, 5 needed modernisation and 2 reconstruction. The aerodromes on the mainland were planned on the basis of the Air Force's appreciation whilst the Army's views were not taken into consideration, as a result of which some of the sites selected were tactically weak to defend.[9] This was to have serious repercussions when hostilities did break out.

Not only was the ground defence of the aerodromes suspect, even the AA defences were later to be found wanting. It was, as a policy, decided that each aerodrome was to be protected by eight heavy and eight light AA guns. This was the ultimate scale of AA defences to be provided at each aerodrome and it was this scale that was considered while deciding on the dispersal areas and splinter-proof pens at aerodromes.[10] But as the ultimate scale was never provided, and most of the aerodromes had no AA guns, the airfields were to become easy targets for the Japanese Air Forces.

The state of radar stations to cover the mainland and Singapore was somewhat better. By December 1941, there were four operational radar stations—Mersing, Tanah Merah Besar (East of Singapore Island), Bukit Churiang (southeastern tip of Johore) and Tanjong Kupang (southwestern tip of Johore) with two still being readied. All these radar stations were seaward looking, with only restricted overland cover, the hinterland being covered by the Observer Corps.[11] It was not a very efficient arrangement but it gave a fair cover for Singapore, if not to mainland Malaya.

Before the outbreak of hostilities, Malaya and Singapore were still weak in air defence. One British unit, the 3rd AA Regiment, and the 11th, 21st and 22nd AA Batteries were all that was available for defence of the naval base and the airfield. It was equipped with 3-inch guns and had two 40-mm. AA guns with some Vickers 2 pounders ex-Royal Navy. The Regiment was to be provided with newer 4.5-inch AA guns but due to delays in provisioning, received 3.7-inch AA guns instead with 16 40 mm. AA Bofors guns. The AA defences were augmented by two HKSRA regiments, the 1st and 2nd AA Regiments, equipped with 3-inch guns. The two regiments also had a LAA battery each.[12]

The AA defences were biased towards the defence of the naval base in the north of Singapore, the four island airfields, the docks, with the harbour and the causeway linking Singapore with Johore Bahru. There was no AA unit in Malaya. In August 1941, the Malaya Command projected the requirement of four LAA regiments, followed up by a request for two Heavy AA regiments for the field force.[13]

The situation somewhat improved with the arrival of some GL Mark I radar sets, and more importantly, the 1st Indian Heavy AA regiment in August 1941.[14] Raised in October 1940, it was the first Indian AA regiment and was still undergoing training when it was ordered to move to Singapore. The composition of the regiment was a mix of British officers and NCOs, Punjabi VCOs and gunners, with a LAA Battery of Madrassis.

The regiment had sailed onboard H.T. *Empress of Japan*. The second ocean liner named *Empress of Japan*, built in 1929-30 in Scotland for Canadian Pacific Steamships and regularly traversed the trans-Pacific route between the west coast of Canada and the Far East until 1942 when it was re-fitted as a troop ship upon the outbreak of the Second World War. The journey had largely been uneventful barring the fact that a gunner, AAA/15672 Gunner that Abdul Ghani, died on board. Sadly, the cause of death is not known and all that remains is the record at the Singapore Memorial.[15]

The regiment, with a total strength of 12 British officers, 9 VCOs, 1,250 IOR and 111 followers, reached Singapore on 5 August 1941. H.T. *Empress of Japan* was received by the Commander-in-Chief, Far East, though he was not to receive the regiment, however, needed it was to beef up the anti-aircraft defences, but to welcome the Maharaja of Patiala who was travelling to Malaya to visit the Indian State Forces.[16]

The regiment disembarked the same day, moving to a camp at Chick Abu near Nee Soon. The war diary of the regiment records that Gunner Abdul Ghani was buried in the Mohammedan troops' cemetery on the 6th of August even as the regiment got down to issuing bedding that 'commenced in the darkness' and setting up the camp in a rubber plantation, under canvas. The regimental war diary also mentions two letters, received from HQ Movement Control and the OC Troops and master of the ship, complimenting the regiment on its discipline and for having left the ship in good state. As the congratulatory letter mentioned, the ship's master 'had never seen troop decks left in a better condition.'[17]

The conditions at the camp were, however, not so good. The state of water supply and the latrines were far from satisfactory with the regiment camped under canvas. The Brigade 'Q' staff visited the regiment but the details of the improvements carried out after the inspection, if any, are not known. The regiment soon got on with a training regime and familiarization with its new equipment. One of the initial problems was the shortage of MT and trained drivers.[18] Though the regiment took over 5 Ford Station wagons and 6 Ford 3-ton lorries from the local ordnance depot, it did not have any trained drivers and had to borrow them from the 4/9 Hyderabad Battalion of the Indian State Forces.

The more important task was to take over the AA guns and the operational tasks. As the various sections and troops of the regiment got deployed, it was handed over a mix of AA guns. The 2nd Indian HAA Battery took over four 3-inch AA gun site from the 13th AA Battery, HKSRA, at the air base while the 3rd Indian HAA Battery received 3 AA guns at Bukit Ayeb for training purposes and the 4th Indian HAA Battery took over the Naval Base (East) position with 4 inch Naval guns. Two sections each of the 2nd and 4th HAA Batteries were deployed at the respective positions.

The light AA guns for the 1st Indian LAA Battery were provided by 14th LAA Battery, HKSRA, which gave two semi-mobile 40-mm. Bofors AA guns on loan. As the regimental gunners were not familiar with the newly taken over guns, training was carried out with the help of HKSRA which also provided instructors for telephone duties and height finding.

Still facing a paucity of trained drivers, the regiment attached 20 of its drivers with the 7th Mixed Reinforcement Camp, Bidadari, for training. Even so, it had to resort to employing Malayan syces as drivers to tide over the deficiency.[19] While the gun drill and associated training was going on, the regiment participated in an AAD exercise conducted by the AD Command, Singapore, in September 1941. It was primarily for HAA batteries and involved cooperation with the searchlight batteries. The regiment strength was augmented by the arrival of a draft of 30 British officers and 88 men of other ranks who joined the regiment on 1 October 1941. Twenty-four of the newly arrived officers were much needed subalterns who could be employed at troop/section level. The officers' strength was satisfactory at last as the regiment had managed with twelve officers till date. The arrival of the draft brought in an additional problem of language and Urdu classes for all newly arrived officers were soon organized by the regiment. In addition, these officers were made to undergo a course in heavy AA gunnery that was conducted at Changi camp.

September-end saw the regiment take over additional tasks with the 2nd HAA Battery taking over Blakang Mati gun site from 12th HAA Battery, HKSRA, with two of its sections deployed in the position. The regiment was settling down though concerns remained about the level of technical proficiency the Indian gunners had. With the training going on under regimental and battery arrangements, it was important to give the gunners laying practice on a live target but the problem was of availability of special aircraft for this purpose. As the demands for air co-operation for the training of AA crews multiplied manifold with the arrival of new anti-aircraft units, there were simply not enough aircraft to meet the requirement. A way out was to combine AA defence exercises with other forms of training in order to make full use of flying hours. Nevertheless, this resulted in many flying hours of operational units, which were all too precious for their own training, being expended on this work, and yet the AA Defence Commander's requirements were never approached.

The arrival of the Indian AA regiment may have improved the state of AA defences in Singapore but overall, especially in mainland Malaya, it was far from satisfactory. There was an additional need of LAA artillery to be provisioned for the field formations and units of III Indian Corps which was tasked the defence of the Mainland. Owing to the paucity of resources, all that was allotted to III Indian Corps was the 16th LAA Battery, HKSRA.[20]

The anti-aircraft artillery was earlier under the Commander, Singapore Fortress, but with the increase in the number of anti-aircraft units and the extension to air defence cover to mainland Malaya, all anti-aircraft artillery assets were brought directly under the command of GOC Malaya, except for purposes of administration, and were reorganized under the Commander, Anti-Aircraft Artillery, into two brigades, one for Singapore and one for North Malaya.[21] This reorganization was still an ongoing process though HQ AD Defences had been established at Singapore with the command of the anti-aircraft artillery in North Malaya being delegated to the Commander, III Indian Corps.

As anti-aircraft artillery units were inadequate, the airfields in Malaya had been allotted a small number of HAA guns manned by HKSRA: Alor Star and Kota Bharu had four 3-inch HAA guns each from the 9th/1st HKSRA Regiment and Sungei Patni had eight static 3.7-inch with two 3-inch guns at Gong Kedah from the 13th/2nd HKSRA Regiment.[22] Unlike the 16th LAA Battery which was controlled by III Indian Corps, the guns at the airfields were controlled directly by the HQ AA Defences at Singapore.

In addition to the dedicated anti-aircraft artillery units, the Indian State Forces' detachments were used to provide air defence and a number of RAF assets were protected by them, manning static naval AA guns.[23] Never having handled and fired these guns, these troops were hastily given a 'crash course' in firing of the AA guns and it was hoped that they would be able to put a decent enough fight.

Trying to do the best with the available resources, the Malaya Command got down to coordinating the AA defences; and the training of troops. Related to this was the training to counter probable Japanese gas attacks as it was assessed that the Imperial Japanese Air Force may resort to aerial spraying of poisonous chemicals including mustard gas. As there were not enough AA guns, use of Light Machine Gun (LMG) was the common resort, especially for the infantry.[24] Sadly, these were merely cosmetic measures, designed more to hope for and delude than protect.

The limited resources, and even more worryingly, the limited understanding of AA defences can be seen in the orders issued by the GOC 9th Indian Division from Kuala Lumpur in a memorandum on the nature of training to be carried out by the 8th and 22nd Infantry Brigades. Emphasis was to be laid during training on countering probable Japanese chemical warfare techniques, and rightly so, but all that was asked was to use the LMG and rifles against planes flying below 1,500 feet as it was felt that 'fire of LMG and rifles against planes doing ground strafing would be effective'.[25]

The priorities laid down in the training memorandum were no less interesting though, as it was quite emphatic in stressing the importance of protection of British Officers' Mess and the barracks of British soldiers. Besides the reliance on LMG and small arms, the other problem was that the training was more in name than in reality. The shortage of ammunition for training was so severe that use of white flags and 'rattlers' were resorted to during practices. Ammunition for AA guns was received only in October 1941 but even then, in absence of tracer ammunition, no practice firing of AA guns could be carried out.

Even if the tracers had been available, it was only the HAA guns that could have been trained as the shortage of ammunition existed in case of 40-mm. AA guns. Such was the state of AA defences of Malaya in December 1941—a mixed bag of troops and weapons with a rather abysmal state of training. By end-November 1941, there were growing signs of Japanese activities, and reconnaissance planes were sighted off Malaya and Singapore.

On 1 December, Order 'Seaview' was received by the regiment. All leaves were cancelled and final preparations were made—preparation at the Air Base, Blakang Mati, and Naval Base for 3-inch guns to fire; one round per gun was fired to check the guns. At the same time, the 4th Indian HAA Battery was ordered to man the guns at Beting Kusa and the 1st Indian LAA Battery was asked to man eight gun positions at the Selatar air base and eight positions at Alexandra and Normanton. Guns for these new tasks were collected from the local ordnance depot. The regiment was also asked to reconnoiter gun positions at the Krangi Oil Depot and Bukit Temah Supply Depot.[26]

With the threat of Japanese attack being imminent, a new order, 'Raffles', was issued on 6 December that required all gun sites to be at 15 minutes readiness. All leaves were cancelled and the 4th Indian HAA Battery was issued a warning order for move to Kalantan in Malaya. By evening, at 1927 hours, 'permission was granted to open fire on any aircraft not established as friendly'.

That was changed within an hour to 'Jap aircraft not friendly and will be engaged.'[27] All these warnings and orders proved to be an exercise in futility as the AA defences were taken by surprise when the Japanese Air Forces raided Singapore for the first time on 8 December. The same day as Japan invaded Malaya.

The Japanese launched their offensive on 8 December 1941 with the 25th Army carrying out an amphibious assault on the northern coast of Malaya at Kota Bharu and attacking western Malaya through Siam (Thailand). Moving rapidly, the Japanese achieved a handsome victory at Jitra, capturing northern Malaya by first week of January. This was followed by capture of Kuala Lumpur on 11 January 1942. With the Allied Air Forces knocked out of the skies, Japan had complete mastery of the skies.

The Imperial Japanese Navy's 22nd Air Flotilla with 110 aircraft, operating out of three air bases near Saigon, took part in the initial attacks on Malaya. It included the 22nd Genzan, Mihoro, and Kanoya Air Groups equipped with 33 Type 96 Mitsubishi G3M1 Nell bombers. The Air Flotilla also had 25 Type 96 Mitsubishi A5M4 Claude fighters. The Japanese Army's component was the 3rd Air Corps and three Air Combat Groups of the 5th Air Corps with 354 IJAAS first line aircraft that included Nakajima Ki-27 Nate, Nakajima Ki-43 Oscar, Mitsubishi Ki-51 Sonia fighters and the Kawasaki Ki-48 Lily, Mitsubishi Ki-21 Sally, Mitsubishi Ki-30 Ann bombers. In addition, Mitsubishi Ki-15 Babs, Mitsubishi Ki-46 Dinah reconnaissance planes were used in the campaign. Most Japanese pilots had combat experience and were all very well trained.

The AA action in the mainland battle was of piecemeal nature and was a one-sided affair with the Japanese Air Forces overwhelming the AA defences. The action at Kuantan was illustrative of the odds faced by the ad hoc AA defences. On 9 December, at 1100 hours, it was attacked by the Japanese twice, with 27 planes each time. The RAF was caught on the ground, losing three aircraft destroyed and five damaged. All that the 'A' Company of the 5th/11th Sikh Regiment, deployed for the defence of airfield, had was LMGs and small arms, and yet they tried to ward off the raiders, firing at them for over two hours but with a foregone conclusion. This was repeated time and again at almost all the airfields in Malaya. All that could be said of the anti-aircraft defences on the mainland was that 'the anti-aircraft defences of the aerodromes everywhere got plenty of practice against bombing aircraft'.[28]

The performance of the Japanese aircraft and their pilots, especially with regard to high-level bombing, came as a surprise to the Allied commanders. The Imperial Japanese Army Aviation first tried to gain control of the air by aerial exterminating actions, then 'considered' support for the ground forces and carried out strategic bombardment as the last stage of its operations. The aerial extermination action involved complete destruction of the enemy aircraft in the air and on ground. While the fighters drew the RAF fighters into combat, the bombers flew some distance back from their fighters and waited till the RAF fighters, due to lack of fuel, were forced to land. Then, the Japanese bombers attacked before the RAF fighters had time to refuel and take to the air again, making them sitting ducks for the Japanese planes. The Japanese bombers in the formation released their bombs simultaneously, on signal from the leader. To cause heavier attrition, Japanese fighters carried out strafing.[29] In all, the Japanese raids were quite effective.

Faced with increasing gaps in the AA defences, HQ AA Defences moved 14th/3rd HKSRA Regiment to the mainland. The 4th/1st Indian HAA Regiment was also kept on standby, but with increasing Japanese attacks on the island, it was not ultimately sent and was held back in Singapore.

As the Allies fell back, GOC Malaya decided to withdraw all troops back to Singapore. In the fog of the battle, contact with forward troops was lost and a series of rearguard actions were converging on Johore Bahru and the causeway linking it to Singapore island. The AA defence of the causeway was now of vital importance. The elements of three HAA and three LAA batteries—11th/3rd, 6th/1st HKSRA and the 4th Indian HAA Batteries and the 16th/3rd, 89th/35th and 144th/35th LAA

Batteries provided the AA cover with a total of about 60 guns.[30] The Japanese, however, failed to exploit the situation to their advantage and held back air attacks on the causeway for a short period, allowing the Allied forces to withdraw with the last element evacuating on 30 January 1942. The AA guns at Johore were also retrieved. The causeway was blown up on 31 January, bringing an end to the fighting on the mainland.

Next was Singapore.

SINGAPORE

Much before the last Allied troops had withdrawn to the island, the Indian AA batteries had already faced the Japanese air raids. Going back to December, the first air raid had come as a rude shock to the garrison. The Japanese formation was detected by a radar station in Mersing, Malaya, almost an hour before they reached Singapore. Three Brewster Buffalo fighters of No. 453 Squadron RAAF were on standby at RAF Sembawang but were denied permission to intercept the Japanese as it was feared that the anti-aircraft batteries would fire on the friendly fighters, despite Vigors being an experienced night fighter in the Battle of Britain.[31]

The anti-aircraft guns immediately opened as the bombers began their attack at 0430 hours but no aircraft was shot down. A total of 17 Japanese aircraft were in the formation, nine of which flew over the targets without releasing their bombs to draw the searchlights and anti-aircraft guns away from the other group. They were flying at 12,000 feet, while the second formation was at 4,000 feet. The war diary of 1 Indian HAA Regiment notes,[32] '19 Jap planes raided Singapore island. They were engaged by AIR BASE and NAVAL BASE EAST. Bombs fell near no. 3 site at SELATAR. A and B troops fired at planes out of range.'

The bombing of Singapore was carried out by 17 G3M Nell bombers of Mihoro Air Group (Mihoro Kaigun Kōkūtai), Imperial Japanese Navy, flying from Thu Dau Mot in southern Indo-China and was timed to coincide with the landing by

Japanese forces on Kota Bharu, Malaya. A total of 65 bombers, 34 bombers of Genzan Air Group (Genzan Kaigun Kōkūtai) and 31 bombers of Mihoro Air Group, had been assigned for the raid and the six squadrons from both air groups had taken off from southern Indo-China on the night of 7 December 1941. However, bad weather caused most of the formations to become separated and after several failed attempts to regroup, the Genzan Air Group aborted the mission with only seventeen G3M bombers of Mihoro Air Group manging to reach Singapore. The original list of targets included the Sembawang naval base and Keppel harbour but with reduced numbers, the raid was directed towards Tengah and Selatar air bases. It failed to cause any serious damage except damaging three Bristol Blenheim bombers of No. 34 Squadron RAF. More serious damage was caused in the town where some two hundred casualties were caused by the bombs that fell in the congested area. A number of bombs also fell on Raffles Place where 61 people were killed and more than 700 were injured. Most of the casualties were troops of the 2nd /2nd Gurkha Rifles, 11th Indian Infantry Division.[33]

The diary of Sergeant Ron T. Buntain of Royal Australian Air Force, then stationed at Sembawang airfield also mentions the raid:[34]

> Monday, 8 December 1941—Was awakened by siren at Naval base at 0345 hours. Air raid warning yellow was given over amplifier system. 453 Squadron told to report to hangar immediately, while on the way to hangar we saw a formation of [Japanese] bombers which had been picked up in the searchlight beams. All told 17 bombers. Anti-aircraft guns went into action. Saw flashes of tracer bullets bursting around the kites [aircraft].

Though the bombing caused only minor damage to the airfields, it stunned the British as they 'hardly expected the Japanese to have any very long-range aircraft'[35] and were under the mistaken belief that Japan's air forces were not capable of striking Singapore from airfields more than 600 miles away in Indo-China. A shock it may have been but many still believed that Singapore would be able to defend itself. As one of the British soldiers recalls:[36]

> Of course we were not too alarmed about the whole thing. Of course, it did give us a fright, I mean, first time experiencing air-raids. But we all believed, I suppose, that Singapore was an impregnable fortress and Singapore would not fall and all that, blah blah blah. So we thought at least the forces that were here would be able to defend Singapore.

There was a respite from further air raids as the Japanese Air Forces focussed on the mainland, in support of their 25th Army operations. Meanwhile, readjustment of AA defences continued at Singapore with new gun positions being reconnoitred and prepared. This was partly due to the pattern of the air raid, with the bombers staying above 20,000 feet and the strafing of the airfields done from high level, that had exposed the gaps in the AA defences with the Japs staying well above the effective ceiling of the AA guns. In order to better utilise the 3-inch guns, it was decided to send them to the airfields but this involved making new emplacements for the static guns held by 1st Indian HAA Battery. But only one such emplacement could be completed in time.

On 11 December, the 3rd HAA Battery was ordered to organise 'Half Battery' into two 40-mm. troops and was issued with nine 40-mm. guns to occupy Alexandra and Normanton sites.[37] This was yet another change for the gunners as they were now required to man LAA guns whilst having been earlier equipped with HAA guns. It was not only the AA troops that were being shuffled around as the 40-mm. guns manned by Royal Navy at the naval base were placed under 1st Indian LAA Battery. While it may have been advisable to centralise the command and control at a site, these changes had, more often than not, an adverse effect on the operational preparedness of the troops, and was to take a toll in the days ahead.

As news trickled in of the Japanese advance in Malaya, the confidence that the Allies could defend Singapore start to waver. What was more worrying was the loss of the much heralded

British warships HMS *Prince of Wales* and HMS *Repulse* that were sunk by the Japanese on 10 December 1941. With this, the Japanese effectively demonstrated their mastery of the sea and the air.

The next raid occurred on the night of 16/17 December 1941 by two Japanese Ki-21s on RAF Tengah.[38] It failed to cause any damage.

There were sporadic night-time hit-and-run air raids on Singapore during December with the AA defences being organised and reorganised. The 13th HAA Battery of HKSRA came under command of the 1st LAA Battery at Kallang as did the 40 mm. guns at Buona Vista. The AA guns of 1st Indian LAA Battery (part of the 13th AA Regiment, HKSRA) at Sembawang shot down a Jap reconnaissance aircraft on 23 December.

New positions were prepared for 40 mm. guns at the Tengah airfield and simultaneously four guns were moved from Tengah to the naval base. The new positions were just about ready when the next air raid came in on 29 December. About 16 Jap planes came in and were engaged by the guns of 1st Indian Heavy AA Regiment but with no results. This was followed by air raid by about 12 planes on 30 December and 17 planes on 31 December. The targets were the air bases, the aerodrome installations, naval base, government offices, and other installations. This time around, the airfield at Tengah was also attacked where the 40 mm. LAA guns tried to engage the Japanese aircraft with no results.

January 1942 saw an increase in air raids by the Imperial Japanese Naval Air Force with an air raid almost every day. The early raids were directed against the aerodromes with Tengah receiving special attention. The native labour at Tengah ran away after the very first serious raid, on 1 January, and that made runway repair difficult due to inadequate labour.[39] With flying effort reduced, an additional burden came on to the AA defences to counter the Jap planes.

During the second half of January 1942, the Japanese Air Forces concentrated their attacks chiefly on the aerodromes on Singapore Island. As the withdrawal from the mainland had restricted the Allied Air Forces' operations from four main aerodromes of the island, three of which—Tengah, Sembawang and Seletar—were sited on the northern side of the island and were, therefore, likely to be usable for a short time only, it was decided to maintain in Singapore only a fighter strength of 8 Hurricanes reinforced by the remaining Buffaloes.[40] With no RAF complements to bank upon, this put further strain on the already hard pressed AA defences.

While the AA defences were biased towards defence of military installations, viz., the naval and air bases, the task of protecting the supply depots and administrative installations, was now becoming equally critical as it was unlikely that any further reinforcements would be forthcoming and the garrison was to rely only on the available stocks. The main administrative installations included the base supply and ordnance depots at Alexandra, the main military supply dumps at Bukit Timah and Thomson roads, ammunition magazines at Alexandra and Changi and the petrol & oil depot at Bukit Temah. Troops of 1st Indian HAA Regiment were deployed at all these installations for their air defence.

The AA defences may have been beefed up with the arrival of additional AA Regiments on 13 January but the withdrawal of forces from the mainland had deprived the AA defences of the help of the warning system which had been carefully built up. Their efficiency was thereby reduced, though the GL sets proved to be of some value.

In the latter half of January, the Japanese Air Forces concentrated their attacks chiefly on the aerodromes on Singapore Island. The Tengah airfield came in for a special treatment from the Japanese Air Forces and was testing the AA defences including the troops of 1st Indian Light AA Battery. It was attacked on the 6th, 8th, 12th, 13th and 14th January with the naval base being the target on the 15th and on the 16th, aerodromes and the docks. Though the raid on the 12th was by over 120 planes, 17 January was a particularly bad day because of its ferocity—attacks on aerodromes were carried out by about 80 escorted bombers, and, under their cover, low flying fighters slipped in and attacked Sembawang and Seletar. Sgt Buntain's diary records the following:[41]

Saturday, 17 January 1942—What a day. Air raid 1030 hrs, a formation of 27 Jap bombers accompanied by Navy Zero fighters came over drome from the direction of the naval base. The Japs were in V formation until our drome was in sight, then they lined out covering the full width of drome. The first bombs landed about the centre of the runway and right across the hangars. . . . One bomb hit our hangar and set fire to a plane in the hangar and boy did the bullets make a terrible noise. The size of bombs ranged from 120 lb. antipersonnel (Daisy Cutters) to 500 lb. general purpose and two men were killed and many injured. Kites in the dispersal bay were set on fire and quite a number of the boys were buried in the shelter trenches. . . .

On several days the civilian casualties ran into several hundreds. The attacks were usually carried out in daylight by formations of 27 to 54 bomber aircraft escorted by fighters in perfect formation despite the AA fire. The maximum number of aircraft which attacked the Singapore area in any one day was 127. As most of the raids were carried out by bombers flying at altitudes over 20,000 feet, they were well beyond the effective range of the 3-inch AA guns which formed one-third of the heavy AA defences. Only the forty 3.7-inch HAA guns offered some deterrence to these high flying Jap planes whilst the 40-mm. Bofors were effective against most of the low level attacks except the few surprise raids.

The raids may have been carried out from heights above 20,000 feet, and the accuracy of bombing was good. The report on the Air Operations in Malaya mentions:[42]

The qualities of the Japanese Air Force came as a complete surprise—in numbers, performance and quality of equipment, training and experience of its personnel, and in its mobility. Its fighters displayed unexpected all-round qualities. They and the Japanese medium bombers had ranges of 1,500 to 1,600. miles which enabled them to operate from bases out of our reach. Their normal operational height was 20,000-24,000 feet where *they were immune from any of our AA gun defences* (*emphasis added*).

Nevertheless, casualties were inflicted on the air formations, though these were difficult to assess, and some of the formations were broken up. The near-immunity with which the Japanese planes could come over Singapore and target any installation was now beginning to affect the morale.[43]

> Well, it wasn't a very nice feeling to. . . You know in those days, we thought that the British were so mighty that nothing could . . . no one could . . . touch them. It was a nasty feeling suddenly, to see these planes coming 27 or 9 at a time sailing through the sky, with not a bit of opposition. And just dropping bombs as they wanted and going off. That was a nasty feeling, and sort of hurt to one's pride that this could be done to one.

The frequent air raids were a cause of concern for another reason—ammunition. There had been heavy expenditure of light anti-aircraft ammunition in the series of battles. As the stocks in these categories had never been very satisfactory, their rapid depletion was now giving rise to some anxiety.

With the fighting going on rather badly on the mainland, and the island repeatedly being targeted from air, General Percival tried to reorient and reenergise the defences. The work to develop new gun sites, along with the massive task of digging, mining and construction of emplacements took off. The one good news was the arrival of reinforcements including one heavy and one light anti-aircraft regiment of the Royal Artillery, giving a much needed impetus to the AA defences.

Such was the state of defences when the Japanese launched the next phase of their operations for the capture of Singapore.

Heavy bombardment in the north sectors of Singapore started on the evening of 8 February 1942, followed by assault by the Japanese on the north west coast. All communications were disrupted due to shelling. The pulling back of the Australian Infantry Brigade opened up a gap in the defences, giving the Japanese easy passage. On 9 February, as the Tengah airfield fell to the Japanese, the 1st Indian LAA Battery deployed at the airfield was amongst the first troops on the island to suffer heavy losses.[44]

The next day, Japanese reconnaissance planes appeared, soon to be followed by bombers which relentlessly attacked positions along the Jurong Road from 0800 hours to dusk. Repeatedly, the Japanese medium bombers, often in a formation of 27 aircraft, came and bombed locations near Jurong Road including the Bukit Timah site. The LAA guns of the 1st Indian LAA battery were hard pressed throughout the day in trying to engage the Japanese planes.

As the Japanese 5th and 18th Divisions continued their advance towards the Singapore perimeter, the troops of the 1st Indian LAA Battery deployed at the supply dumps at Bukit Timah were now in the line of fire, getting involved in hand-to-hand fight with the Japanese. The 40-mm. AA guns were used in a direct firing role, firing over open sights but the loss of Bukit Timah meant that the ammunition for the Bofors was running out now.[45] The gunners took on the Japanese with their small arms, while at times, it was a hand-to-hand fight that the gunners had to resort to.

After the loss of Bukit Timah, the Australian brigades and the British 18th Division was withdrawn to a perimeter which now extended from Normanton to Kallang, as Changi had been isolated by the Japanese. Both locations had elements of Indian AA Regiment—1st LAA Battery at Normanton and the 3rd HAA Battery at Kallang. Kallang came in for special treatment by the Japanese, with over 200 bombs being targeted against it, causing heavy casualties and leaving the entire place in a mess of craters. The HAA guns were under constant attack and it left the 3rd HAA Battery with over 20 dead and many more injured during the day.

The date of 13 February saw the Japanese troops attack Pasir Panjang Ridge, as it provided passage to Alexandra, where there is a hospital, as well as British ammunition and supply depots. The fuel depot at Normanton caught fire as a result of which the troops of the 1st Indian HAA Regiment deployed there fought a day-long battle amidst flames, smoke, shelling and small arms fire, all the while trying to keep their guns in action. This earned its troop commander, Lt John Ashcroft Hopson, a well deserved Military Cross. His citation reads:[46]

This officer displayed great gallantry and devotion to duty on 13 February 1942 when at Normanton, in the face of heavy shell fire and bombing, he continued to organize and keep in action his troop which several times during the day suffered casualties to guns. His disregard for danger and his own personal efforts in replacing or repairing damaged equipment and encouraging his men enabled guns to be kept in action to the last possible moment.

His action was a source of inspiration not only to his men but also to the neighbouring infantry.

The Japanese tried to push back the defensive line held by 1 Malaya Infantry Brigade after crossing the Jurong River but the line held. To reinforce the defences, a detachment of the 3rd Indian HAA Battery was sent to reinforce the 1 Malaya battalion which had been pushed back but was holding on to their defensive line at Asiatic Petroleum Depot. Here, too, the gunners acquitted themselves well, fighting alongside the infantry soldiers.

With the defences fast crumbling all over the Island, General Percival held a conference on 13 February to decide on the future course of action. All his senior commanders brought out the fact that the men were exhausted and disorganised, leaving out any prospects of an effective counter-attack. The next day, close fighting continued with the entire defensive line under attack. The LAA guns had no ammunition, fuel was running out and water supply was getting affected. Fire was razing at the depot and magazine at Alexandra. Faced with a near hopeless situation, Percival decided to surrender, contacting the Japanese on 15 February.

The next day Japanese occupied Singapore. The Battle of the Fortress was over.

All units suffered heavily and the 1st Indian HAA Regiment was no exception. It was severely tested during the battle, its troops fighting isolated battles against all odds; no communications, loss of ammunition and cut off from their headquarters, the gunners acquitted themselves with honour. Lt Colonel John Rowley Williamson, the commanding officer, was honoured with a Distinguished Service Order (DSO), one of the troop commanders won a Military Cross and two gunners, Havildar Sham Lal and Gunner Balbir Singh, were awarded the Indian Distinguished Service Medal (IDSM) but all this came at a heavy price as the regiment lost over 95 men in the one week of fighting with another 200 dying in custody in prisoner of war camps.[47] Of the twenty-four subalterns who had joined the regiment in Singapore on 30 September 1941, six died by April 1943, while in captivity. 2nd Lieutenant Ronald Jack A. Hutchins of the 3rd/1st Indian HAA Regiment was one of them. Son of George and Kathleen Hutchins, he was only 25 when he died.[48]

The 1st Indian HAA Regiment was never raised again, with official records noting it as 'lost in Singapore'. It was a tragic end for a valiant regiment but the story of the AA regiment did not end there. The Japanese, and Mohan Singh of Indian National Army (INA), were keen to recruit the Indian *sipahis* and visited the camps where the Indian jawans were interned. A number of gunners of the 1st Indian HAA Regiment also joined the INA, with most of them serving in the Guerrilla Regiments of the INA. Some of them fought against the 14th Army in Burma; Gunners Amar Singh, Bansi, Bishambar Singh, Brahma Datt, Ganeshi Lal, Mangal Singh, Richpal Singh of 1 Indian Heavy AA Regiment amongst them.[49]

Today, the War Memorial at Singapore is the only testimony to the valour and bravery of the first Indian AA Regiment to be bloodied in battle.

NOTES

1. 'Operations of Malaya Command from 8 December 1941 to 15 February 1942', Supplement to *The London Gazette* of Friday, 20 February 1948, pp. 1250-62.
2. Ibid., pp. 1339-40.
3. Ibid.
4. Benjamin Lai, *Hong Kong 1941-45: First Strike in the Pacific War*, Bloomsbury, 2014 (Kindle Edition).
5. 'Operations of Malaya Command from 8 December 1941 to 15 February 1942', op. cit., pp. 1266-7.
6. 'Operations in Hong Kong from 8 December 1941 to 25 December 1941', Supplement to *The London Gazette* of Thursday, 29 January 1948; and Lai, op. cit.
7. Ibid.

8. 'Operations of Malaya Command from 8 December 1941 to 15 February 1942', op. cit., pp. 1247-9.
9. Ibid.
10. 'Operations of Malaya Command from 8 December 1941 to 15 February 1942', op. cit., pp. 1255-6 and N.W. Routledge, *History of the Royal Regiment of Artillery: Anti-Aircraft Artillery,* Brasseys, London, 1998, pp. 220-1.
11. Ibid.
12. Routledge, op. cit., pp. 220-3.
13. Ibid.
14. War Diary of 1st Indian HAA Regiment for August 1941.
15. Ibid.
16. Ibid.
17. Ibid.
18. Ibid., September 1941.
19. Ibid.
20. Routledge, op. cit., pp. 220-3.
21. 'Operations of Malaya Command from 8 December 1941 to 15 February 1942', op. cit., pp. 1255-6 and Routledge, op. cit., pp. 220-3.
22. Ibid.
23. Bisheshwar Prasad, *Campaigns in South East Asia 1941-42,* Pentagon Press, New Delhi, 2012, and Kaushik Roy, *Battle for Malaya: The Indian Army in Defeat, 1941-1942,* Indian University Press, Indiana, 2019.
24. Kaushik Roy, *Tropical Warfare in the Asia-Pacific Region, 1941-45,* Routledge, London, 2018, and *Sepoys Against the Rising Sun: The Indian Army in the Far East and South-East Asia 1941-45,* Brill, Leiden, 2016.
25. Ibid.
26. War Diary of 1st Indian HAA Regiment for December 1941.
27. Ibid.
28. Alan Warren, *Britain's Greatest Defeat: Singapore 1942,* Hambledon Continuum, London, 2001, p. 75.
29. War Diary of 1st Indian HAA Regiment for December 1941.
30. Routledge, op. cit., pp. 220-1.
31. Brian Farrell, *The Defence and Fall of Singapore,* Monsoon Books, Singapore, 2018 (Kindle Edition).
32. War Diary of 1st Indian HAA Regiment for December 1941.

33. Roy, *Sepoys Against the Rising Sun: The Indian Army in the Far East and South-East Asia 1941-45.*
34. 'The War in Malaya,' Sgt Buntain's Diary, The Warbird's Forum, accessed at https://www.warbirdforum.com/453sq.htm. Accessed on 23 February 2020.
35. Lt Gen A.E. Percival, *War in Malaya,* Orient Longman, Bombay, pp. 109-14.
36. 'The War in Malaya,' Sgt Buntain's Diary, The Warbird's Forum.
37. War Diary of 1st Indian HAA Regiment for December 1941.
38. Ibid.
39. Ibid.
40. Percival, op. cit., pp. 109-14.
41. 'The War in Malaya,' Sgt Buntain's Diary.
42. 'Report on the Air Operations during the Campaigns in Malaya and Netherlands East Indies from 8th December 1941 to 12th March 1942,' Supplement to *The London Gazette* of Friday, 20 February 1948.
43. 'The War in Malaya,' Sgt Buntain's Diary.
44. War Diary of 1st Indian HAA Regiment for December 1941.
45. Routledge, op. cit., p. 224.
46. Citation for Military Cross with Respect to Lt John Ashcroft Hopson accessed at National Archives, UK at https://discovery.nationalarchives.gov.uk on 10 September 2019.
47. Records maintained for Singapore Memorial by Commonwealth War Graves Commission, accessed at https://www.cwgc.org/find/find-cemeteries-and-memorials/2053500/singapore-memorial on 10 September 2019.
48. Ibid.
49. Prem Nath Chopra, *Who's Who of India's Martyrs*, vol. II, Publications Division, Ministry of Information and Broadcasting, Government of India, New Delhi, 2013.

CHAPTER 4

Retreat from Burma, 1941-1942

The situation in the Far East and the peril of an outbreak of war in Europe led to a review of the measures for defence of Burma and, in 1938, an inter-services conference was held in Rangoon that recommended certain urgent measures to be taken in this regard. Based on these recommendations, plans for anti-aircraft defence of Rangoon, the oilfields area, oil refineries at Syriam and other strategic points was prepared and the need for anti-aircraft guns identified. One of the recommendations was the raising of an anti-aircraft regiment but all the decisions were kept in abeyance.

When the Far East Command was established in November 1940, it included Burma and it was not till 15 December 1941 that the responsibility for the defence of Burma was transferred to India under the Commander-in-Chief, India. Burma, at the beginning of the war, was grossly unprepared with inadequate troops, poorly trained and equipped with little, in way of an administrative organization. Placing Burma under the Far East Command did have one positive effect that the 1st Anti-Aircraft Battery, RA, was formed in February 1941 as part of the Rangoon Field Brigade.[1] It was commanded by (then) Major H.T. Hogan and was initially equipped only with 20 light machine guns. It had no anti-aircraft guns.

The battery remained a part of the field brigade until 15 August 1941, when it was detached to form part of the 1st Heavy Anti-Aircraft Regiment, RA, BAF. The regiment had two batteries, the 1st Heavy and the 3rd Light Anti-Aircraft Batteries.[2] As the officers and men of the newly raised batteries were not familiar with the anti-aircraft techniques, some of the officers and men were sent to Singapore to undergo a course on anti-aircraft gunnery.

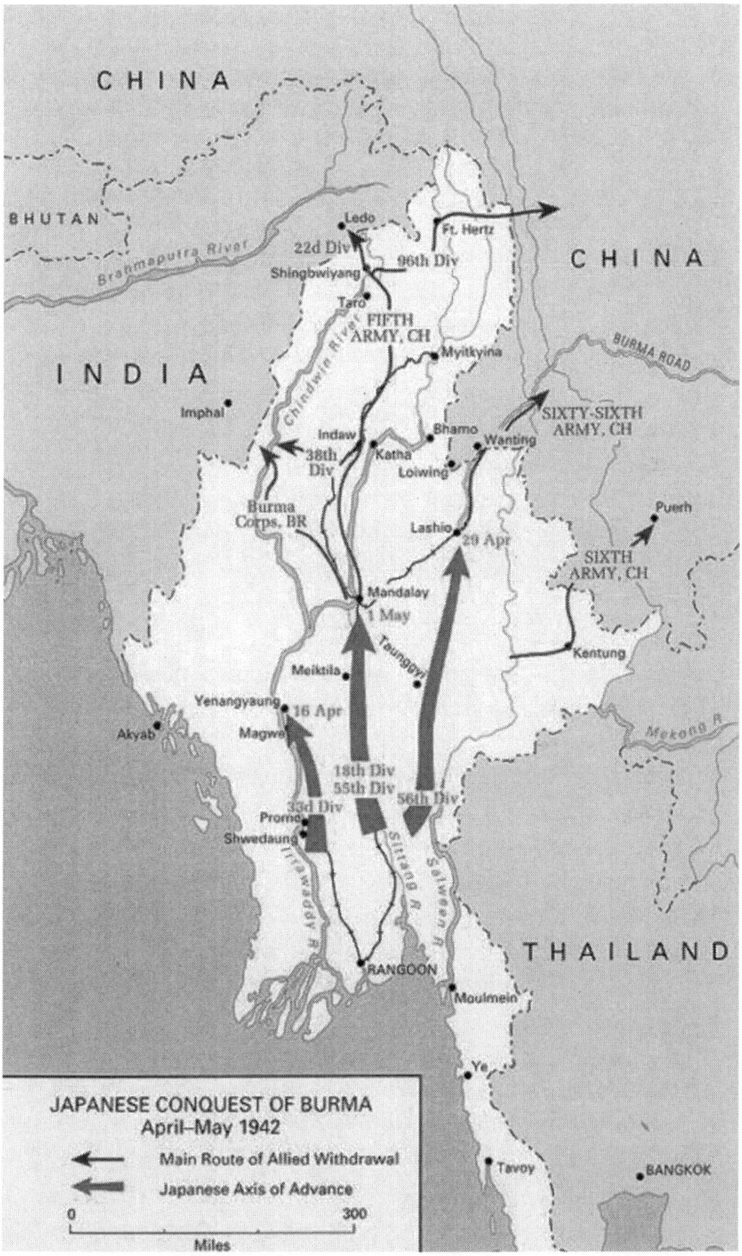

Map 2: Burma 1941-2. *Source:* U.S. Army Center of Military History: Burma, 1942, 7 December 1941-26 May 1942, U.S. Army Center of Military History (Open Domain) SEAC plans for Burma 1944-5.

However, the regiment received its guns only after the outbreak of war with Japan when four 3.7 inch mobile guns were transferred from India for the 1st HAA Battery, RA, BAF. One section of the battery, with two guns, was at Dundeedaw and the other was at Ahlone, covering Rangoon docks. The 3rd LAA Battery, BAF under Major J. Irving, was equipped with eight 40 mm. Bofors Mk I guns and thus had only two troops of four guns each. One troop was at Syriam, protecting the oil refinery and installations, the other was deployed at Mingaladon Aerodrome.[3]

Five days after Japan entered the war, Burma was placed under the commander-in-chief in India with the 18 British Division, then on passage to Mid-east, allotted for the defence of India and Burma. In addition, commitment to send the 17th Indian Division to Iraq was taken back and a special allotment of anti-tank and anti-aircraft guns was promised along with four squadrons of fighter aircraft.[4]

In Burma itself there were at this time two British infantry battalions, two Indian infantry brigades, and eight battalions of Burma Rifles. The artillery in Burma comprised only four mountain batteries and one four-gun 18 pr. battery. There was also the Burma Frontier Force, the equivalent of approximately six battalions, divided into a number of detachments.

The air forces nominally available in Burma at the outbreak of war were one fighter squadron equipped with Brewster Buffaloes and one bomber squadron of Blenheims. One squadron of American Volunteer Group (AVG), an air force manned by American and equipped with P-40 fighters, was also available for the defence of Rangoon.

Realising the inadequacy of the anti-aircraft defences with a partly-equipped 1 HAA Regiment, Burma had prepared detachments of the Burma Auxiliary Force for light automatic defence against low-flying aircraft, each equipped with 10 to 12 .5 Browning machine guns on AA mountings and these were stationed at Mingaladon and Zayatkwin. The detachments were manned entirely by Burmese personnel; mainly from the 12th Burma Rifles. They were of limited deterrent value but did add to the overall anti-aircraft defences. The surveillance and early warning network were similarly antiquated and inadequate. Of the planned network of three chain stations, two Chain Overseas Low (COL) and two Ground Control Interception (GCI), only one COL set was in the country and was initially located at Moulmein but was moved to Rangoon for a better arc of observation.

The first Japanese attack against Burma came on 14 December 1941, against Victoria Point, almost the most southerly point of Burma. It was expected that the Japanese would initiate the battle from there and attack was not even contested with the BAF just giving up that area. It was followed by a small probing raid directed at a police station in southern Tenasserim, which was repulsed.

On 23 December 1941, Rangoon suffered its first bombing raid and its first civilian casualties. It was carried out by a large fleet of 87 bombers supported by 34 Type 97 fighters, from the airfields in Thailand and Indo-China, flew over the Gulf of Martaban towards Rangoon and the Mingaladon airfield. Many of the bombers were twin engine KI21 Mitsubishis each carrying a 2,200 lb. bomb load. The Mingaladon airfield was 15 miles north of Rangoon.[5]

Just prior to the air raid, four static 3.7 inch guns had arrived by sea at Rangoon for the 1st HAA Battery, BAF. The guns had been unloaded and were in the process of being prepared for action when they were caught in the raid. Two of the guns, still in their protective jackets, were lying out in the open and were hit. One was damaged by fire but was put into service, giving three static and four mobile 3.7-inch guns for the defence of Rangoon. The Japanese raid was met by anti-aircraft fire from the seven 3.7-inch guns of the 1st HAA Battery, BAF and the eight 40-mm. guns of the 3rd LAA Battery, BAF. The Japanese raid not only targeted the docks but also the aerodrome at Mingaladon and one bomb struck a Bofors gun of the 3rd LAA Battery, killing one officer, one NCO and five gunners. Three enemy aircraft were claimed shot down but at the cost of one 3.7-inch and one 40-mm. AA gun.[6]

The 3rd Indian LAA Battery had, in the meanwhile, been moved to Calcutta from Malir, arriving in third week of December 1941. Maj C.H.T. MacFetridge had been transferred from the 1st Indian HAA Regiment to command the battery as it was to set sail for Burma. On arrival at Calcutta, MacFetridge deployed his twelve guns in a perimeter around the city with one gun on top of a warehouse as that was the only open area available near the docks. As the battery prepared to embark for Rangoon, MacFetridge organised an attestation parade for the gunners—an oath to loyally serve the crown. The parade was held on top of the warehouse and was administered by the regimental moulvi.[7]

The Battery embarked on a Free French ship on 23 December. The battery commander wanted to deploy his guns on board the ship in case the Japanese air forces attacked them during the voyage but the captain of the ship forbade him from positioning any guns ahead of the bridge in case they were stopped by the Japanese Navy. His fear was that as a merchant vessel, he was forbidden to have guns deployed ahead of the bridge and in case he was stopped by the Japanese Navy, his ship would be seized for violating the norms! No guns were deployed in any case and the Japanese did not attack them or check them during the brief voyage.[8]

As the reinforcements were still en route, the Japanese came back on 25 December to bomb the docks at Rangoon causing severe damage and loss of life.[9] Six days later, the reinforcements despatched from India, the 8th Indian HAA and the 3rd Indian LAA Batteries, reached Rangoon to be greeted by the smell of rotting bodies. Major MacFetridge, recollects:[10]

> On arrival at Rangoon we found the docks deserted because the Japs had bombed the docks on Christmas Day. We found a good assembly area about fifteen miles near Mingaldon airfield. I deployed two troops to defend the airfield, and one troop moved by rail and boat over the river Salween to Moulmein.

The 8th Indian HAA Battery was allocated to the defence of Rangoon and surrounding military and air force installations. The battery had been formed as part of the 2nd Indian Heavy Anti-Aircraft Regiment, which was raised on 1 April 1941. The regiment moved to Calcutta in October 1941, from where the

8th HAA Battery had been sent to Burma. It was commanded by Major R.A. Roberts who was originally a sapper officer, having been commissioned in 3 (Bombay) Fortress Company, Royal Engineers, Auxiliary Force, India, on 17 December 1938. After having served with the Fortress Company, Roberts was transferred to Indian Artillery as a major and was given command of the 8th Indian HAA Battery, Royal Artillery.

The Indian AA batteries were followed by the 8th HAA Battery, RA, that arrived at Rangoon on 16 January 1942 with its four 3-inch mobile AA guns.[11] It was immediately sent to Syriam and relieved the 8th Indian HAA Battery which moved to Martaban, where two guns of the battery had been protecting the ferry to Moulmein since 12 January.

Early in January 1942, HQ 17th Indian Division with one infantry brigade arrived at Moulmein with orders to hold the Japanese till such time Rangoon was reinforced. Moulmein was an important communication centre due to its airfield and, being a route centre, plus the nearby port and railhead of Martaban. In addition to a troop of 3rd Indian LAA Battery, Moulmein also had the 8th Indian HAA Battery with its four 3-inch AA guns.[12]

On 5 January, the Japanese planes attacked Moulmein but suffered no casualties. As the RAF commander did not want to risk any fratricide incident, the AA guns had not been allowed to fire, especially at night, but this order was revoked on 6 January and a procedure for alternate gun night and fighter night was instituted allowing the guns to engage the Japanese planes on the designated 'gun nights'. As per this policy, the LAA guns were supposed to fire a barrage of 24 rounds per engagement at any plane seen or heard over the aerodrome. With no GL sets or searchlights, the HAA guns were to fire using the geographical barrage method but it severely limited their efficacy.[13]

The next time, on 9 January, when the Japanese planes came to bomb Moulmein, they were met by a barrage fire. No damage was done though the Japanese formation was broken and the attack did not go through. Noticing a pattern in the Japanese attacks—a smaller group preceding the main attacking formation—the AA defenders had an apprehension that the Japanese were using the firing of guns to locate the airfield. In order to try and deceive the Japanese about the exact location of Moulmein airfield, the 'C' Troop of the 3rd Indian LAA Battery was shifted to Insein township. The next few raids were broken up by the LAA fire and, though it could not be ascertained if the deception worked or not, the War Diary of 3rd Indian LAA Battery does mention that[14] 'Our guns are forcing the Jap planes to abandon low level attacks by night and are now coming at great height, over 20,000 feet.'

Meanwhile in the south, the Japanese launched their attack against the airfields at Tavoy and Mergui in Tenasserim. Spearheaded by the Japanese 143rd Infantry Regiment of 55th Division, the attack against Tavoy was launched on 18 January with the airfield falling to the Japanese without much resistance while Mergui capitulated without any resistance.

By 20 January 1942, a section of the 8th Indian HAA Battery had deployed to Moulmein where it took up defence of the airfield and came under the command of the 17th Indian Infantry Division.

No sooner had the HAA guns deployed that the aerodrome at Moulmein was attacked on 20 January. The anti-aircraft guns opened up as the Japanese planes came in. Some Blenheims were also in the air at that time and, in the prevailing confusion, the heavy AA guns fired at them (Blenheims). Luckily there was no damage to the RAF aircraft or any casualties. The guns of the LAA troop had also opened up, firing 79 rounds and was duly 'accused' by the RAF Group HQ of having fired at their own aircraft.[15]

On 22 January 1942, the Japanese 55th Division began the main attack westward from Rahaeng in Thailand across the Kawkareik Pass, advancing towards Moulmein, not facing any major opposition. On 24 January, the Japanese carried out the first day light air raid since 25 December, but lost all planes to the RAF. Another raid followed in the afternoon with three planes swooping down on Moulmein airfield. This time two of the guns of the 3rd Indian LAA Battery opened up and scored a hit.[16]

The Japanese attack on Moulmein was swift and effective, forcing the 17th Indian Division to withdraw. On 30 January 1942, at the height of the Japanese attack on Moulmein, the Battery was evacuated across the river to Martaban. About 1500 hours the same day, the guns of LAA troop were near the jetty ready to be loaded on launches for transit to Martaban when Japanese aircraft made a bombing attack. The brigade commander present asked that the guns should remain at Moulmein and his request was granted. It is very doubtful in any case if the guns could have been evacuated, for at 1700 hours Japanese infantry on rafts drifted down the river and landed about 2 miles north of Moulmein.

Second lieutenant Webber of the 3rd LAA Battery assisted 8th Indian HAA Battery in extricating two 3-inch heavy AA guns using his Troop transport to leave Martaban by road to Kyaitko, at the height of the Japanese attack on Moulmein.[17] Taking advantage of the chaos, Japanese soldiers dressed up as Burmese and mingled with the retreating Burmese troops, catching the Indian gunners by surprise, bayoneting many of the LAA gunners.

The gunners fought valiantly in a hand-to-hand combat to save their guns though they failed to do so. Fortunately, all predictors and generators had been evacuated three days earlier. The LAA gunners were confident that they had inflicted many casualties by means of small arms fire and the bayonet. They managed to disable only two of the four guns when they were forced to abandon their guns at the jetty.[18] MacFetridge recollects:[19]

> On the last day of January I decided to pay the troop at Moulmein a visit, as a boost to morale. I thought the Japs were still well to the south in Tenassarim. While I was travelling in a Tiger Moth with the soldiers' pay in a bag, the pilot indicated that he had been recalled to base in Rangoon, telling me that the Japs had overrun Moulmein. So I had four guns on the wrong side of the mighty Salween river. In the confused fighting there I lost all four guns.

Second lieutenant Mehar Dass, the sole Indian officer with the troop, was, however, not willing to let go and as he came to know that the Japanese had left the Bofors unguarded led a team of gunners to try and retrieve his gun, and failing so, to destroy it. He had voluntarily returned to the jetty knowing fully well that he was sacrificing his only chance to escape. But try as he may, he could not get the gun across the river and was pulled on to the last streamer by the withdrawing johnnies. As a number of his gunners were left behind, Mehar Dass dived off the streamer and swam ashore under heavy fire to make one last attempt to save his men. The after action report mentions:[20]

> The conduct of 2/L Mehar Dass & 34578 L/Naik Tawasin was outstanding. The former returned to the guns already captured by the enemy and with the aid of Tp HQ personnel, orderlies, signallers & drivers brought one gun out of action and down to the jetty by means of a 15cwt. lorry. Later he got off the last launch to destroy this gun & assist survivors.
>
> No. 34578 L/Naik Tawasin took the place of the gun havildar, when killed on Jan. 22 with distinction and ability. Later he swam the River Salween complete with rifle and amn.

At that stage it could not be confirmed if Lt Dass survived and was taken prisoner or killed in action. The war diary of the battery records:[21]

> One KCIO (2/Lt Mehar Dass I.A.), one BOR, seven IOR and one follower missing due to enemy action.
>
> Four guns, two FA tractors, one 3 ton lorry, one 15 cwt. lorry. One water lorry, seven rifles and 500 rounds Bofors amn lost due to enemy action.

Years later, in 1945, when one of the gunners, AAA/4105 Sahib Khan, of the battery, who had been taken prisoner at Moulmein managed to return to India, via China, it came to light that a number of gunners were taken prisoner and employed as labour by the Japanese on the Lasho Road. Sahib Khan also confirmed that 2nd Lt Dass was amongst the prisoners but was taken to Rangoon and interned at the Central Jail.[22]

After the fall of Moulmein, the 17th Indian Division was forced into a rear-guard action as it moved to protect the bridge on Sittang River leading to Pegu and Rangoon. It attempted to hold the Bilin river and other fallback lines but, with too few troops, was repeatedly outflanked. The division eventually retreated towards the bridge over the Sittang river in general disorder, delayed by air attacks and Japanese harassment. The 8th Indian HAA Battery and a troop of the 3rd LAA Battery tried to cover the withdrawal across the bridge. The 'C' troop of the 3rd Indian LAA Battery that had lost its guns at Moulmein was hastily reorganised as it took over four guns of its Burma namesake on 17 February 1942, at Mingaladon.[23] Around 19 or 20 February 1942, the 8th Indian HAA and elements of the 3rd Indian LAA Battery were ordered to withdraw across the Sittang to defend the bridge from the west bank of the river. The HAA and LAA guns were withdrawn but the other troops were ordered to hold their positions. This delay in withdrawal by the 17th Indian Division allowed the Japanese to infiltrate across the Sittang, posing a serious threat to the bridge and the risk that it may fall into the Japanese hands. In such circumstances, the GOC ordered the bridge to be blown up. Most of the troops were still on the enemy held side, across the river.

With the Japanese across the river, Pegu was the next key point on the road to India. The 8th Indian HAA Battery and remnants of the 3rd LAA Battery, now left with five guns, moved to Pegu. En route, the two fought against repeated Japanese air attacks; destroying Jap planes—in one day they shot down six of them. The going was, however, tough with a bad road and repeated enemy ambushes with the 3rd Indian LAA Battery losing a 40 mm. gun in one such ambush.[24]

At Pegu, the 8th Indian HAA Battery was deployed to defend the HQ of the 17th Indian Division. The troop of 3rd Indian LAA Battery was in action by 24 February with all guns west of the river, defending the bridge over the river. The Japanese air raids came in from 4 March onwards but were initially repelled by the LAA guns, which shot down a Japanese plane on the first day itself. The next day, Japanese Air Forces came in strength but suffered a loss of three aircraft.[25] 'Formation of seven light bombers and three fighters engaged. Two bombers and one fighter definitely hit and seen falling. This is the most successful engagement till date.'

By the evening, the Japanese infantry had closed in and though the troop managed to withdraw east of the river, one of the Bofor guns got caught in an ambush and was lost with its detachment.

The 3rd Indian LAA battery was now redeployed, one troop

at Rangoon to cover the disembarkation of armour and another going to the bridge at Hlegu, with the troop at Hlegu made up of remnants of B and C troops with five guns and named as 'BC' troop.

At this stage, General Harold Alexander took over the overall command in Burma. On 13 March 1942, Burma Corps was created under Lt General William Slim to take control of the scattered British, Indian and local troops retreating through Burma. The main components of this force were two infantry divisions, 17th Indian Division and 1st Burma Division with the recently- arrived 7th Armoured Brigade Group.

The overall plan was to create an east-west front across the Irrawaddy and Sittang Rivers ahead of the Japanese advance in the hope that the Chinese divisions under General Stilwell would move south to hold the eastern front around Toungoo. The routes open from Rangoon were the one via Pegu, Toungoo and Meiktila in the east and via Prome and Magwe in the west. The plan was to concentrate the 17th Indian Division in the Prome area and the 1st Burma Division was to move up on the eastern axis and attempt a link up with the Chinese at Toungoo. The task of the AA batteries was to defend the western axis up to Prome, and the airfield at Magwe. Two squadrons of RAF and one of AVG were now located at Magwe, flown in from Rangoon.[26] As the Burcorps was in no position to execute both the move North and defend Rangoon, it was decided that Rangoon would be evacuated.

On 7 March 1942, with the port at Rangoon and oil terminal blown up, the Burma Army evacuated Rangoon. The static 3.7-inch HAA gun positions were demolished and the remaining 3-inch and 3.7-inch guns were taken to Hlegu. The 8th Indian HAA Battery, having moved to Hlegu, deployed at its new location where it claimed two out of six Japanese aircraft attempting to attack the withdrawing troops. The 3rd Indian and 3rd BAF LAA Batteries followed, having sent the predictors to India by sea.[27]

Both the routes had been cut by the Japanese, with a series of blocks. One of the first roadblocks faced by a column of the 17th Indian Division was at Taukkyan where the Japanese Air Forces were also very active. In support of the troops was every anti-aircraft gun in Burma. Within a 2 sq. mile area were heavy and light guns from the 1st Heavy AA Regiment, BAF, the 8th Heavy AA Battery, RA, the 8th Indian Heavy AA Battery and the 3rd Indian Light Battery.[28] The 8th Indian HAA Battery claimed two Japanese planes shot down of the six engaged, while the 3rd Indian LAA Battery claimed one. The Japanese planes used to try and locate the gun positions from their fire, and then specifically target them. This practice of Japanese air forces was noticed later on also but in one of the first instances on 6 March, the troop of the 3rd Indian LAA Battery came in for a special attention from the Japanese.[29]

> During the afternoon a formation of twelve light bombers made a determined attack from 8,000 ft on No. 2 gun and dropped twelve 250 lb. bombs. One fell within 40 yds., two within 30 yds. and one within 10 yds. The sand bag wall saved the gun det. It will be noted that this was the only gun that had not changed its position since Feb 28; it had probably been located by the enemy recce planes.

After several unsuccessful attempts to clear the Japanese, it was found on the morning of 8 March that the Japanese had pulled out owing to a tactical lapse by the Japanese Commander, leaving the road clear and enabling the Allied troops to withdraw. The formation succeeded in getting past Taukkyan without any incident.

The 8th Indian HAA Battery withdrew to Tharrawaddy during 9 and 10 March and from where it later joined the other anti-aircraft units in defending Shwedaung, Prome, Allanmyo and Magwe. By the time the 1st Burma Corps or '*Burcorps*' came into being on 19 March 1942, the 8th Indian HAA Battery was assigned as Lines of Communication Troops.[30]

The batteries moved in a leapfrog manner through ambushes and flanking attacks with the pressure increasing in the rear as the Japanese brought in more troops through Rangoon. As gaps existed between the brigade and battalion columns, the AA batteries moved from one infantry locality to another, with positions inside brigade 'boxes.' With bad roads, the problem of vehicle shortage got aggravated to an extent that 8th Indian HAA

Battery had to raid dumps of unserviceable vehicles to try and salvage some vehicles to remain mobile. As all Allied aircraft at Magwe had been lost after a massive air raid by Japanese 5th Air Division employing 250 aircraft on 22 March, the onus of defending the retreating columns was on AA artillery alone.[31]

In April 1942, most of the surviving anti-aircraft guns available to the British were concentrated around the vital Irrawaddy bridges at Mandalay. With Mandalay being the hub, it was important to protect the bridges from the Japanese, hence the last fortnight of April had three HAA and one LAA batteries, along with the remnants of the 3rd Indian LAA Battery deployed around Mandalay. The five batteries though had a grand total of only nine heavy and 14 LAA guns between them. This proved to be more than a match for the Japanese who failed to destroy the bridges, allowing the Allies to cross the river. By 30 April, the rear-guard of the 17 Indian Division had crossed the Irrawaddy at Sagaing.

As the Japanese moved with unexpected speed up the Sittang valley, cutting off and annihilating the Chinese 200th Division, the link-up with the Chinese on the Eastern front was no longer feasible; moreover, it totally exposed the eastern flank. This necessitated the defence of Mandalay, and of two bridges over the Myitnge and Irrawady rivers, lest they fall to the Japanese. To do so, the 1st Burma Division was ordered to hold the Yenangyaung oilfields while the 17th Indian Division held a defensive position south-east of the oilfields, also preparing for a counter-attack to clear the eastern route. The wide dispersion meant the AA batteries again had to spread out thin with the 8th Indian HAA Battery at Magwe moving its 3-inch guns by barges upstream to Sagaing while its vehicles took the circuitous route through Meiktila.[32] One of the troops of 3rd Indian LAA Battery moved to Mandalay where two 40 mm. guns were placed on each of the two bridges. The 8th Indian HAA Battery was later replaced by 1st HAA Battery, BAF and 8th HAA Battery, RA, at Magwe. Meanwhile, Prome was subjected to a heavy air raid, in which the 3rd Indian LAA Battery shot down five Japanese planes.[33]

At Yenangyaung, the 1st Burma Division, facing a series of

Japanese attacks on 10-17 April, had to withdraw with the orders to blow up the oilfields being issued by General Slim on 15 April. In the battle at Yenangyaung, the troop of the 3rd Indian LAA Battery was overrun and all its guns captured. With its 40 mm. guns destroyed, Subedar Nur Khan, one of the VCOs of 3rd LAA Battery, personally manned a Bren Gun and engaged the Japanese planes as they came into attack the seemingly defenceless gunners, hitting one plane. Nur Khan had been wounded at Taukkyan but had carried on serving with his troop. At Yenangyaung, he personally destroyed all the wireless sets and vehicles before withdrawing. He was awarded the Indian Order of Merit for his act of gallantry.[34]

The battery commander, Major C.H.T. MacFetridge, had to once again gather his men and guns, spread over Central Burma, and try to keep up with the withdrawing columns. After the loss of Yenangyaung, the 17th Indian Division held the Japanese advance at Kyaukse, holding back the Japanese advance to Mandalay. The Burma Corps crossed the Irrawaddy river, north of Mandalay with the 17 Indian Division, 7th Armoured Brigade and a Chinese division, making a wide circuit to Shwebo, Ye-U and Kaduma. The 1st Burma Division continued moving north astride the Chindwin. Both the columns were to converge at Shwegyin and Kaduma, from where the withdrawal to India was to begin.

From here onwards the going got rougher, with bad roads and tracks. The jungle gave some protection from air attacks but the stretch along the clearings exposed the retreating columns to Japanese air. The move was made more difficult for the AA batteries due to frequent breakdowns and accidents as they moved in small parties, towing guns with whatever transport they could muster. The entire route was first required to be reconnoitred; before undertaking the move, making it an arduous task. MacFetridge, the OC of 3rd LAA Battery, went over the entire route till Shwegyin, to check the feasibility of moving the AA guns. In his words:[35]

We were ordered to move to Shwegyin on the east bank of the River Chindwin. From the jetty here river boats ran to Kalewa on the west bank and from there a track led to Assam and India. I did a recce and was so appalled at the state of the road to Shwegyin, I thought I would be very lucky to get any guns to Shwegyin. I imagined this would be the 'Dunkirk' of the Burma Army. Guns would have to be jettisoned at Shwegyin. On return from my recce, I told Brigadier Welshman at General Slim's HQ what I had seen. I was taken in to see General Slim. After listening to my description of the road, Slim summoned his CRE and G1 Ops. I cannot but think that Slim took the swift action, because the road was vastly improved during the next few days.

Despite the improvement, the last stretch of the road was atrocious, along the edge of a *chaung*. I only managed to get four guns through. I was told to break all rules of deploying AA guns and put them close together like field.

Before Shwegyin, there was another task for the AA batteries. The wounded soldiers needed to be evacuated and the selected landing ground to do so was at Shwebo. To protect the 8th Indian HAA and 1st BAF HAA Batteries, a troop of the 3rd Indian LAA Battery was deployed at the landing ground. It was soon enough targeted by a massed attack by 27 Japanese planes but was defended well by the AA gunners.[36]

Shwebo was the end of the road for the heavy 3.7-inch AA guns as the track ahead did not allow them to be taken any further. The 1st Heavy Anti-Aircraft Battery, BAF disabled its 3.7-inch guns at Shwebo and the BAF gunners then joined the long march to India.[37] This was the end of the road for the AA batteries of BAF as the 3rd Light AA Battery, BAF, had earlier lost all its guns. Throughout the campaign, the 1st Heavy Anti-Aircraft Regiment, BAF, had claimed 19 Japanese aircraft destroyed of which four were claimed by the 3rd Light AA Battery, BAF.

The 8th Indian HAA Battery was able to take along its guns as it withdrew towards Kalewa but was forced to destroy the guns en route at Ye-U. With all guns, transport and equipment destroyed, the unit joined in the withdrawal from Shwegyin and Kalewa on 10 May, crossing the Chindwin river.

The British 8th HAA Battery and a troop of 3rd Indian LAA Battery were sent to Monywa on the Irrawaddy where the 40-mm. Bofors guns came in handy in direct firing role against Japanese barrages. Next day, they were ordered back to Ye-U. The British 8th HAA Battery could, however, move up to Kaduma only, about 80 km short of Shwegyin, where they had to destroy their guns as the track did not allow them to be taken further ahead. Now, the only AA guns left in Burma were the six 40-mm. Bofors of 3rd Indian LAA Battery. Their problems were, however, not yet over. During the move to Shwegyin, two of the Bofors skidded off the track as the battery negotiated a steep cliff; as a result, the Battery reached Shwegyin on 6 May with only four 40-mm. AA guns.[38]

As the retreating columns tried to reach Shwegyin and Kalewa, the two launch sites on the banks of river Chindwin, the Japanese tried to encircle the Burma Corps troops and cut off its withdrawal routes. The launch sites, about 10 km apart, were held by the infantry as an 'bridgehead.' The Shwegyin basin was about 400 × 1,000 yds. with a steep escarpment along the perimeter with a track running across it to the jetty. The basin was held by two companies of $1^{st}/9^{th}$ Royal Jats, reinforced by $1^{st}/7^{th}$ Gurkhas. Three 40-mm. Bofors were deployed along the track in ground defence role with the fourth sited at the entrance of the basin in an air defence role.[39]

The situation is recounted by MacFetridge:[40]

> By the morning of 10 May three of my guns were deployed at Shwegyin. I realised that in the darkness that the Japs had started attacking the troops holding the hills overlooking the ferry boats evacuating the army. The surviving companies of 1st Royal Jats were exchanging fire with the enemy. They were to be relieved by the 7th Gurkhas, an amalgamation of the $1^{st}/7^{th}$ and $3^{rd}/7^{th}$ Gurkha Rifles. I asked their CO if they had any mortars, and he did not know. The two COs agreed that the 7th Gurkhas should not relieve the Jats but reinforce them. I told the COs that I had only three Bofors deployed that could support them, but I was a mountain gunner who had fought on the North-West Frontier, which they seemed to find reassuring. The CO of the Jats, Godley, placed his HQ by the guns and used them as match winners.

As the Indian and British troops passed through the basin, the deployed infantry and the Bofors covered their move but on 8 May, the Japanese planes came in strength, finding it to be an easy target. The LAA battery shot down two Jap planes but the air raids had their effect as the Burmese crew of the ferries refused to move down south of Kalewa, and only in the night at that. As a result, two infantry brigades with 25-pounder guns and tank crews had to halt about 15 km short of the basin, waiting to embark. The situation became critical as the Japanese troops of 213th Regiment landed about 12 km below Shwegyin on 9 May, on both sides of the river. The Japanese advanced northwards, coming in contact with the perimeter deployed for the protection of the basin. With no. 25-pounders, the defenders had only the 40-mm. Bofors for support, and the AA gunners rose to the challenge as they provided covering fire to the Gurkhas who repeatedly counter-attacked the Japanese to hold on to the perimeter. As the last ferry left, the Gurkhas again attacked the Japanese who had secured a position on the crest overlooking the basin. The 40-mm. firing over open sights provided fire support. The Bofors successfully knocked off a Japanese field gun that had come up on the crest.[41]

> The Japs managed to get one of their mountain guns forward to shell the ferry site. My Bofors gunners picked up this gun arriving and as it was being brought into action it was destroyed by my Bofors firing over open sights. In the course of the battle the Gurkhas were committed to scale the hill overlooking the ferry site and reinforce the Jats. They wore yellow screens as they would have done on the frontier. This enabled my gunners to see where they were and to fire in their support without hitting them.

With no ferries left, the GOC, the 17th Infantry Division decided to evacuate the basin, and destroy all the equipment. The defenders would then move on foot to Kalewa. Avoiding Shwegyin, the rear columns of the 17th Infantry Division were to follow, abandoning their tanks and vehicles.[42]

> We were told to withdraw that night personally by General Cowan, GOC 17 Division. As he drove off in his jeep after leaving us, his ADC was wounded by small arms fire. When the time came to start the withdrawal, I was puzzled as to how Colonel Godley would pass his orders to his forward troops. He asked for two pairs of signallers, and he then gave both of them orders, which they were to give to the company

commanders. The senior VCO told them to repeat the orders back. I doubted they would ever find the forward company commanders, but they did. The withdrawal was a model of its kind. I still had plenty of ammo, and devised a fire plan to bring down the heaviest fire possible to cover the withdrawal.

In reserve were the 1st/4th Gurkhas, under Lentaigne, he reckoned that in the chaung leading to the ferry there were at least three thousand troops awaiting evacuation, and they owed their survival to the actions of the Bofors manned by the Punjabi Mussalman gunners, and the two battalions; a fitting end to the longest withdrawal in the history of the British Army.'

As darkness fell, the infantry withdrew from the cliff via the stream and were supported by the anti-aircraft guns of the 3rd Indian LAA Battery firing off all remaining ammunition at 'automatic', a total of almost a thousand rounds. The firing was so intense that the one serviceable barrel of the A4 gun burst towards the end. The troop then destroyed the guns by removing the breech blocks and breaking the sights. The lorries were destroyed by a pickaxe by Jemadar Abit Ali. The gunners then took their packs and withdrew on the road eastwards.[43] Amongst the gunners of 3rd Indian LAA Battery who were recognised for their bravery at Shwegyin were Gunner Suda Rang who was awarded the IDSM and Jemadar Abit Ali and Havildar Sardar Khan who were awarded Mention in Despatches.[44]

At 8:00 p.m., the last concentration of fire done, the guns and lorries disabled, the LAA gunners formed up under the battery commander. As MacFetridge recalls:[45] 'At the end I ordered the detachments to withdraw. And merely said, "Walk back to India".'

The battery moved to Kalewa in the night, reaching the crossing point early in the morning. They were almost the last to cross the Chindwin, and were certainly the last to have fought the Japanese.

The battle at Shwegyin was, in many ways, the Dunkirk of the East. A gallant retrograde operation carried out over a thousand miles against great odds, ending in a withdrawal across a river. Just two Indian infantry battalions and a light AA battery was all that fought at Shwegyin and yet it demonstrated what can be achieved by courage and determination. The 3rd Indian LAA Battery had lost all but four of its guns during its retreat from Rangoon to Shwegyin. It had suffered heavy casualties, had been on the move, supporting the infantry all throughout the long move back. Fittingly, the last engagement with the Japanese was by the 3rd Indian LAA Battery. It was credited with 22 Japanese aircraft in five months and was awarded two Military Crosses, one Distinguished Conduct Medal, one India Order of Merit, three Indian Distinguished Service Medals and five Mention in Desptaches. Its war diary proudly records:[46]

Battery fought from Moulmein to Shwegyin, a distance of 1,000 miles. 22 hostile aircraft were confirmed as hit:

At Moulmein	1
At Sittang	1
At Pegu	6
At Tankhkyan	1
At Taukkyi	1
At Prome	3
At Pyalo	1
At Tamo-Magwe-Yenangyang	6
At Taunmingyi	1

Other planes were undoubtedly hit.

In addition, one enemy field gun and one mortar were knocked out. *16 guns lost. All transport lost. But due to no fault of the battery.*

The battery tried to save as much as possible of the battery records, taking them loaded on a bullock cart but not all documents could be saved as the path became more torturous towards the end. What documents did survive do mention that the British OR compared very badly as marchers with Indian OR during the withdrawal—except those in the battery commander's party.

The other Indian AA battery, viz., the 8th Indian Heavy AA Battery claimed 8 Japanese aircraft. A total of 19 aircraft were claimed to have been shot down by the 1st Burma Auxiliary Force (BAF) AA Regiment, 15 by HAA and four by the LAA. Four aircraft were shot down by the 8th AA Battery, RA, but it had suffered 70 casualties during the campaign.[47] These figures do not represent the total Japanese losses from ground fire as they were known to be heavy.

The Burma Corps reached Imphal in Manipur just before the monsoon broke in May 1942. The ad hoc Burma Corps HQ was disbanded and IV Corps HQ, which had recently arrived in India, took over the front. With this action came to end the first phase of Indian AA Artillery in combat. The loss of 1st Indian HAA Regiment, and the guns of the 3rd Indian LAA Battery notwithstanding, the Indian AA regiments and batteries had done creditably. With the Japanese at the doors of India, they would now get involved in the defence of the 'homeland'.

NOTES

1. N.W. Routledge, *History of the Royal Regiment of Artillery: Anti-Aircraft Artillery*, Brassys, London, 1998, p. 230.
2. War Diary Rangoon Field Brigade, WO 172/640; War Diary 1st HAA Regiment, BAF, WO 172/749.
3. WO 172/749.
4. 'Operations in Burma from 15 December 1941 to 20 May 1942', Despatch submitted to the Secretary of State for War on the 14 July 1942, by General Sir Archibald P. Wavell, G.CB., C.M.G., M.C., A.D.C., Commander-in-Chief, India, *The London Gazette*, no. 38228 dated the 5 March 1948.
5. Routledge, op. cit., p. 231.
6. WO 172/749 and C. Shores, B. Cull and Yasuho Izawa, *Bloody Shambles*, vol. 1, Grub Street Press, London, 2013.
7. As recounted by Maj. C.H.T. MacFetridge, *Oral History* at Imperial War Museum, London, accessed at https://www.iwm.org.uk/collections/item/object/80020185 on 15 July 2019.
8. Ibid.
9. 'Air Operations in Burma and Bay of Bengal from 1 January to 22 May 1942', Supplement to *The London Gazette* of Tuesday, 5 March 1948.
10. As recounted by Maj C.H.T. MacFetridge, *Oral History* at Imperial War Museum, London.
11. War Diary of 8th HAA Battery, RA, for December 1941.
12. Routledge, op. cit., pp. 231-2.
13. File No. 601/2497/WD, War Diary of 3rd Indian LAA Battery for January 1942.

14. Ibid.
15. Ibid.
16. Ibid.
17. Ibid.
18. Ibid.
19. As recounted by Maj C.H.T. MacFetridge, *Oral History* at Imperial War Museum.
20. File No. 601/2497/WD.
21. Ibid.
22. Ibid.
23. Ibid.
24. Ibid.
25. Ibid.
26. Routledge, op. cit., pp. 230-2.
27. File No. 601/2497/WD.
28. Routledge, op. cit., pp. 230-2 and File No. 601/2497/WD.
29. File No. 601/2497/WD.
30. War diary, 8th Indian HAA Battery, WO 172/835.
31. Routledge, op. cit., pp. 230-2.
32. War diary, 8th Indian HAA Battery, WO 172/835.
33. File No. 601/2497/WD.
34. Records at National Archives, UK at https://discovery.nationalarchives.gov.uk accessed on 10 September 2019.
35. Julian Thompson, *Forgotten Voices of Burma*, in Association with The Imperial War Museum, London, Ebury Press, 2009, pp. 38-45.
36. War Diary, 8th Indian HAA Battery, WO 172/835; War Diary 1st HAA Regiment, BAF, WO 172/749.
37. War Diary, 1st HAA Regiment, BAF, WO 172/749.
38. File No. 601/2497/WD.
39. Ibid.
40. Thompson, op. cit., p 38.
41. Ibid.
42. Ibid.
43. File No. 601/2497/WD.
44. *The London Gazette*, Supplement 35929, dated 9 March 1943, p. 1118.
45. As recounted by Maj C.H.T. MacFetridge, *Oral History* at Imperial War Museum.
46. File No. 601/2497/WD.
47. Routledge, op. cit., p. 235.

CHAPTER 5

Defending the Homeland

The first AA battery to be deployed in India was the 8th AA Battery, RA, when it had arrived in November 1926 and sent to NWFP. But it was deployed at the wrong place when the war broke out in the Far East, as it was facing Afghanistan, looking out for any threat from the Russians. With its eight guns and the additional commitment to provide a detachment to Aden, even otherwise it was hardly in a position to defend the 'homeland'. The forming of the Indian AA Artillery in late 1940 was first attempt to have a credible AA artillery element to defend the 'homeland' but before the first battery could be formed, the threat to East India had manifested itself and it meant that deployment of AA batteries could no longer be delayed even if they were not yet fully trained or suitably equipped.

The first Indian anti-aircraft unit to be deployed for the defence of the homeland was the 1st Indian HAA Battery, to provide anti-aircraft defence to Digboi oil refinery and other installations in Assam. The battery was commanded by Major Charles H.T. MacFetridge, RA, who had been transferred from the mountain artillery after having undergone an anti-aircraft artillery course in the UK. During the reconnaissance, the initial sites selected for the AA guns were the lawn tennis court and the site earmarked for the bungalow of the manager of the Digboi oil refinery.[1] These sites were, obviously enough, not used and the guns were deployed at alternate sites. Before the 3.7-inch heavy guns were received by the battery, it was issued with five 18 pounder Mk.I guns, for they were the only guns available in India—the other AA guns available were being used for training of the recruits at Malir Cantt.[2]

The initial days of settling in were tough for there was no

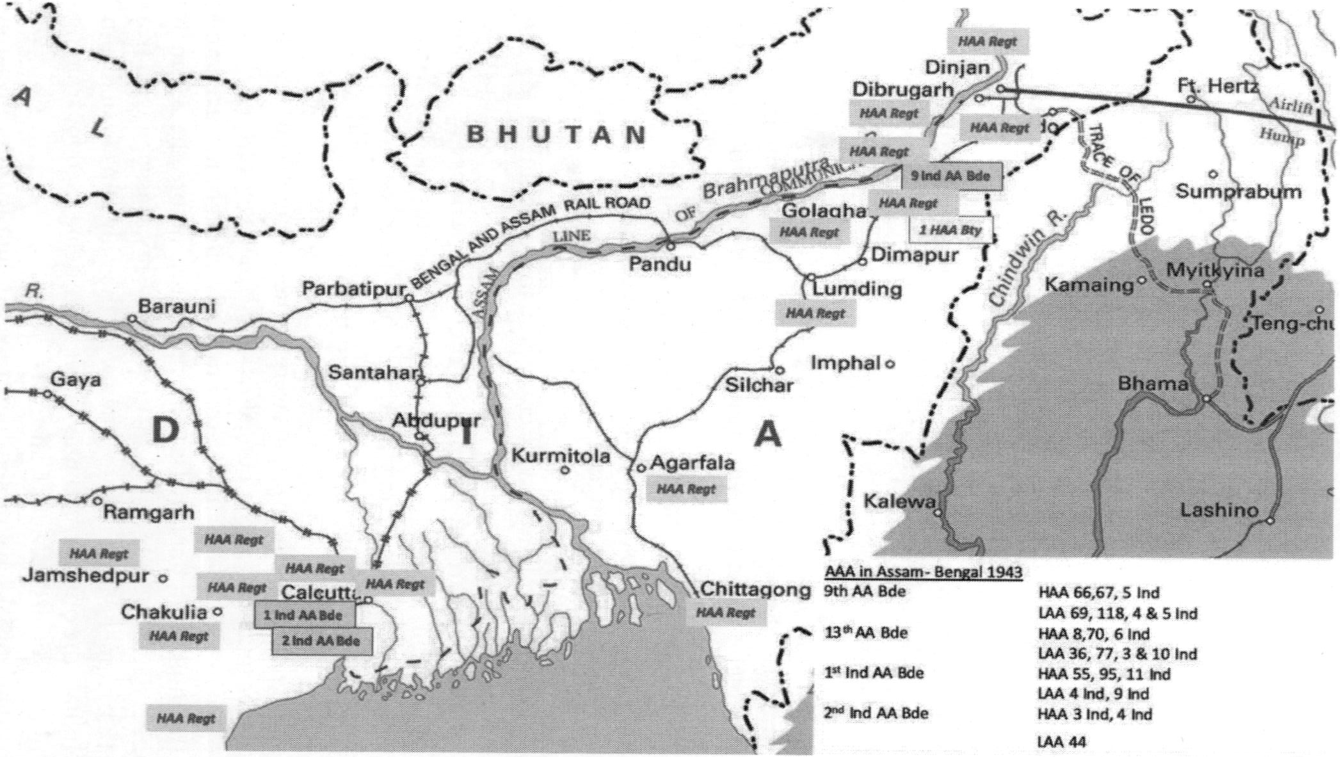

Map 3: Map showing deployment of AA Brigades and Regiments, April 1943. *Source:* Author's collection.

support from the regimental headquarters, over 1,500 km away and it often issued directions that were hard to comply with; like taking precautions against the sun and digging of slit trenches when the heavy rainfall in the area made these orders all very useless. The seemingly routine administrative issues like sanction for the treatment of gunners at the local civil hospital often took most of the time, leaving MacFetridge hard pressed for time to look after the operational matters.

As the Japanese threatened the British outposts in the Far East, the battery's parent regiment, 1st Indian HAA Regiment, was earmarked for Malaya Command with 1st Indian HAA Battery to remain deployed at Digboi.[3] As all available 'fit' officers were to accompany the regiment, MacFetridge was relieved by one of the regimental officers who had been found medically unfit for overseas service and recalled back to Malir Cantt. But rather than being posted back to the 1St Indian HAA Regiment, he was transferred to the 3rd Indian LAA battery of the 1st Indian LA Regiment—still in the process of training.[4]

The period till May 1942 was otherwise uneventful when the battery received a GL Set Mark II for early warning as there were no surveillance radars covering the area. The log of the GL set notes that it was switched on daily from 0300 hours to 1900 hours with four hours given for maintenance from 0600 hours to 1000 hours. As the weather was hot and humid, the radar operators often struggled to keep the radar serviceable.[5] Thankfully, there were no raids, as yet, to test its efficacy.

The raids were to come after the Japanese had finished with the task in Malaya where the battle had raged on from December 1941. The developments in the East were keenly followed and even as the battles of Singapore and Burma were being fought, the Indian Army was in the process of gearing up for the new challenges. The Army was expanding, reorganising and modernising, with major developments taking place in anti-aircraft artillery.

At the outbreak of war, artillery was represented on the General Staff by a Major General Royal Artillery (MGRA) as adviser. Initially, anti-aircraft defence was a general staff responsibility with no dedicated officer looking after the subject at the Eastern and Southern Army Headquarters but the need for technical and procedural knowledge in planning air defence caused the two Army Headquarters to appoint a Brigadier Anti-Aircraft (BAA). The Major General, Royal Artillery, at the General Headquarters looked after the anti-aircraft matters using the BAA channel. With an increasing inventory of anti-aircraft artillery, a deputy director, staff duties (Brigadier), for anti-aircraft duties was appointed at the General Headquarters in January 1942. By the end of the year, the Royal Artillery Staff was reorganized as a separate directorate.[6]

Owing to the raising of further anti-aircraft units being planned including light AA regiments and induction of newer guns, a need was felt to have separate training centres for heavy and light AA artillery. Accordingly, No. 2 AATC was formed at Deolali on 1 January 1942 with the batteries of AATC, Karachi, which had hitherto fore trained personnel for light anti-aircraft transferred with their officers and NCOs to start the new centre.[7] With this, AATC, Karachi, was renamed as 'No. 1 AATC' and was now meant for training of heavy AA gunners only.

As the raising of the planned anti-aircraft regiments would take some time, it was decided to move some British anti-aircraft regiments to India to meet the immediate needs and two regiments, the 67th HAA and 24th LAA Regiments, RA, arrived in India in mid-January 1942 from Middle East and the United Kingdom, respectively.[8] The raising of new Indian anti-aircraft artillery units continued in the new year, with the 3rd Indian LAA Regiment being formed at Karachi on 1 February 1942. The first Indian AA Brigade Headquarters was formed at Calcutta on 26 February 1942 with Brigadier G.P. Thomas, OBE, as the first brigade commander. This was followed by the first Gun Operations Room (GOR) that was formed at Calcutta a month later, on 1 April 1942.[9] Though the anti-aircraft defences were dependent on the radar network for early warning, like everything else, radars were a rare commodity in 1942. At this stage, there was only one early warning radar, at Matherapur, about 80 km southeast of Calcutta; none existed in Assam.[10] The early warning system was rudimentary, but the raising of the GOR was at least a start in putting in place a control and reporting system for providing the early warning of raids to the anti-aircraft units, and also to control the fire of AA guns.

As the arrival of British AA regiments only partly met the requirement, it was decided to convert select infantry battalions to hasten the availability of the required number of anti-aircraft units as these battalions could provide trained soldiers who had to be trained only on anti-aircraft techniques. A total of seven anti-aircraft units were so formed; three HAA and four LAA regiments, starting with the conversion of the 6th/1st Punjab Regiment into the 6th Punjab Light AA Regiment in February 1942.[11] These units and the dates of conversion are:

Date	*Infantry Battalion*	*Converted to*
15 February 1942	6th/1st Punjab	6th Punjab LAA Regiment
	8th/7th Rajput	7th Rajput HAA Regiment
1 April 1942	7th/7th Rajput	8th Rajput HAA Regiment
	8th/11th Sikh	8th Sikh LAA Regiment
	9th/11th Sikh	7th Sikh LAA Regiment
1 May 1942	9th/6th Rajputana Rifles	9th Rajputana LAA Regiment
1 June 1942	14th/8th Punjab	9th Punjab HAA Regiment

Source: History of Army Air Defence, AAD Association, New Delhi, 2008, p. 22.

These anti-aircraft units retained their infantry regimental nomenclature with the units even using their infantry insignia with the artillery badge. These regiments added to the growing number of anti-aircraft units but the shortfall still persisted as a result of which additional units were moved from UK, including HQ 9 AA Brigade, RA, that arrived in Bombay (now Mumbai) in March 1942, and was sent to Assam where a number of airfields and logistics bases were coming up.[12]

After his return from Java in February 1942, General Wavell had begun the process to build a base for defence of India which could also serve as a start point for a new offensive. Dimapur, on the Manipur front, was chosen for this purpose and it was scaled to cater for the requirements of three divisions. Simultaneous to the logistic buildup, work began on construction of nearly 200 airfields in the Ganges plains, Bengal and Assam to provide fighter, bomber and air transport support. The air support for the Burma Corps was also provided from these airfields. In this process, the Americans helped with engineering support, machinery and labour, as also in carrying out air strikes against Japanese positions in Burma.

By the end of the year, there were 88 airfields ready to use, each with at least one 1,000 yds. all-weather runway, spread as far ahead as Digboi. As all these airfields were within the range of Japanese Air Forces, Wavell asked for 66 fighter squadrons and 1,350 AA guns. With the demands being made by all commands and fronts for additional units, there was great pressure on anti-aircraft command but considering the criticality of the situation, nine anti-aircraft regiments and HQ 13th AA Brigade, RA, were moved from UK and reached India in March-April 1942.[13] But the pressure to raise an Indian Artillery to expand and form new AA regiments remained.

The proposal of using only Madrassi men for raising anti-aircraft regiments was now implemented in right earnest and to cater to the increased demand, a No. 3 AATC was decided to be formed. As the first step, training staff was formed at Mehgaon in central India and a nucleus was attached to No. 2 AATC, to be moved to Mehgaon as soon as the accommodation was ready. The proposed British/Madrassi expansion scheme envisaged using the same British NCOs as instructors at the training centres and in the new units to be formed with Madrassi recruits. They brought in a new system of training in which the recruits received two months of basic military training and were then posted in sufficient numbers to form two heavy and two light anti-aircraft batteries a month. These batteries remained at the same location (at the AATC), undergoing technical training with the officers and NCOs of their designated units. After four months, the batteries were grouped as regiments, ready to take on a static operational role.

As all the raisings were to be done along similar lines, the No. 1 AATC was reorganised on the lines of No. 3 AATC and now catered for training of Madrassi recruits only. With two of the three AATCs catering for Madrassi recruits, pre-war class recruits were to be, henceforth, trained only at Deolali.

The other developments taking place in the Army that would affect the anti-aircraft defences were the reorganisation of the Armies in India and the raising of a Women's Auxiliary Corps (India). The Indian Army had retained its old peace time organisation which divided the subcontinent into separate commands till 1942. These commands were redesignated as armies in April 1942 with each army made responsible for security and operations within its area. The Eastern Army thus became responsible for the Assam/Burma front, including the anti-aircraft defences.

The second development was the formation of the Women's Auxiliary Corps (India) on 9 April 1942.[14] Its objective was to assign some of the jobs pertaining to military service to women and, thereby, release men for service in the field. The Corps was constituted by means of an Ordinance No. XIII of 1942 and was first formed to provide suitable personnel for service in the army only. Amongst the many duties assigned to the WAC(I)s, were plotting duties in AAOR centres and also direction finding.

As these developments were taking place, Japan carried out the first air raid against a major city in India. The Burma Corps was still in the process of reeling back and had not yet reached Imphal, yet the reverse in Burma had brought the Japanese threat direct to India. The air threat posed by the Imperial Japanese Air Forces to the mainland was real and could no longer be ignored. The first Japanese air raid in South Asia came not against India but Ceylon with Colombo and Trincomalee as the targets.[15] In the first air raid on 5 April 1942, the Japanese bombarded Colombo and on 9 April, 9 Japanese planes raided Trincomalee.[16] According to Winston Churchill, war-time Prime Minister of Britain, the Japanese air raid over Ceylon was the most dangerous moment of the Second World War and the one that caused him most alarm.[17]

Owing to a lack of radar cover, the raiders, all eighty four bombers, accompanied by thirty-six Zero fighters, were not detected till such time they were almost overhead, and the anti-aircraft defences were largely ineffective in the absence of any warning. The No. 990 Balloon Squadron which had reached Colombo in March 1942, and was deployed at Trincomalee and Colombo added to the overall anti-aircraft defences. Though no incident of a Japanese plane getting entangled in their cables ever occurred, the balloons did managed to keep the Japanese planes away.

The raid on Vizag (now Visakhapatnam) and Madras were a similar experience that followed soon after the Colombo air raid. On 6 April, the Japanese raided Madras and Vizag. A description of the raid that was published in *The Hindu* recounts:[18]

A Japanese aircraft-carrier following the two American ships, loaded with ammunition for the allied forces stationed in Burma, sent two of its fighter planes around 8 a.m. on April 6 sensing danger, the sirens blew and the people of the town who were already given a drill in case of eventualities ran for shelter. The planes after a few rounds of scouting disappeared, and the Vizagites were relieved.

But as the saying goes "never underestimate an enemy", they were back again around 1 p.m., and this time they came in two-plus-three attack formation and started pounding the coast. But the moment they tried to enter the port area, the American ships stationed there opened up their famous 'Bofors' anti-aircraft guns in staccato. The retaliatory fire, combined with the natural protection of the hills on all three sides, did not allow the planes a peep at the inner harbour, and they had to abort their mission.

In the course of strafing, they dropped three bombs near the port area.

One of them hit a concrete pipe five feet in diameter, in which five workers had taken shelter and all of them were killed.

The second one fell on a cement storage site and the third hit the thermal power station that was maintained by the Government of Madras. There were no further casualties and no other buildings were damaged in the raid.

They made a final attempt around 6 p.m. and had to return almost empty handed except for a bomb, which fell on one of the targeted ships, *Marine Maller*, which mercifully did not explode. If it had exploded than that would have set up a chain reaction as the ship was laden with, 2,000 tonnes of explosives and Vizag would have gone into the annals of World War II history, as the second Pearl Harbour.

Thankfully, there were no more air raids as Japan was focussed on its operations in Burma, giving some respite and time to build up the AA defences. But as Yasmin Khan in her book *India at War: The Subcontinent and the Second World War*, mentions, the threat of Japanese air raids caused panic amongst the common people and gave rise to rumours which only further fuelled the panic.[19]

> Sulochna Sisadhari, a schoolgirl from a prosperous family, remembered the fore brooding atmosphere in the Vizagapatnam of her childhood, the blackouts, rations and panic, which now bore some resemblance to the British home front: 'We were scared of war as we were afraid of bombing by enemy countries. We used to live near the sea coast of Visakhapatnam. Defence services used to have coastal batteries, anti-aircraft guns and searchlights. Every week they used to practice firing into the sea. We used to watch as were close by and got scared. We had training in air raid precautions.'

The reaction to the air raids was pure panic at times. With hindsight, the reactions may seem irrational but the fear in the civil populace was real. The civil population left the towns in hordes and though this may be attributed to lack of knowledge what was surprising was the reaction of some of the government agencies. The High Court of Madras was moved to Coimbatore, the inspector general of police to Vellore, the board of revenue to Salem and the secretariat to Ooty, with only a skeleton staff remaining at Fort St. George, Madras. A platoon of Malabar Police shot the animals of Madras Zoo so that they would not try to escape—as it was feared that wild animals running amok could lead to more panic. Three lions, six lionesses, four tigers, eight leopards, four bears and a black panther were shot down within an hour. The elephant was spared as there were not enough people available to bury it.[20]

As the news of the Japanese air raids spread to other parts of India, so did the panic. Jamshedpur saw an exodus of workers, much before the air raid on Calcutta later in December. Even Bombay on the west coast was not immune to rumours and a number of rich business families sent their women and children to Gujarat; much like what happened in London during the blitz. Six extra trains were run every day from Bombay to cope with the additional load. But all this without any real raid coming in. It took some time before the panic subsided and people started returning. In this, the anti-aircraft units played a big role as their presence went a long way in allaying the fears of the populace.

In May 1942, when the Burma Corps reached back and Japan had overrun Burma gaining control of forward airfields to launch raids against India, the anti-aircraft defences were far from ready. HQ Eastern Command had only one AA brigade—1st Indian AA Brigade located at Calcutta with five British AA Regiments, 67th and 95th HAA, 24th, 28th and 59th LAA Regiments, eight batteries of 1st, 3rd and 7th Indian HAA Regiments and the 1st Indian LAA Regiment. Gun positions were occupied or being prepared for the industrial areas around Calcutta, the airfields in Bengal, Digboi airfield, oil installations in Assam and the logistic base in Dimapur.[21]

The gun positions around the airfields and other installations were not the typical dugouts made by the gunners, themselves, but were prepared by the Military Engineering Service (MES). It was not unusual for the work being delayed or not up to the required standard as the war diary of a member of the Indian AA Brigade notes:[22] 'Lt Col Gale, OC 66 Heavy A Regt, RA, conferred with the Brigade Cdr regarding the deployment of his Regiment in the Asansol area with special reference to GE Asansol's delay in completing gun positions and accommodation.'

The construction of gun was not merely that of a 'gun pit' but also the entire infrastructure required to support the gun detachment. Often, it involved constructing a small 'link road' from the nearby main road, clearing of any hutments or even a village, clearing of trees and construction of '*bashas*' or temporary accommodation. Payment of compensation to the villagers was invariably involved and, due to these related expenses, the MES carried out the task. The expenditure of a single site could be as much as Rs. 1,50,000.[23]

As the construction of gun sites often got delayed, it adversely affected the overall preparedness. Another factor was the quality of new gunners who were pushed forward of training centres to meet the demand of new batteries. With the pressure to train a growing number of recruits, the state and standard of training varied, with the new regiments understandably made up, largely, of inexperienced gunners.

Most of the regiments were short of equipment and this shortage of guns and other equipment was to continue during the year and most of 1943. As the regiments were equipped for a static role, they had no integral transport to move their guns and heavy equipment. To make it worse, only one road transporter existed in all of Bengal, capable of transporting the 3.7 inch static AA guns.[24] It was a challenge for the staff to move the guns over long distances and ensure deployment at the right time and right place.

With the spreading network of airfields and other installations in Assam and Bengal, there was a need to expand the strength of AA defences to three AA brigades—two for Calcutta and one for the new airfields and installations in Assam. With this also came the plan for an effective radar network. As the airfields came into being, India and Ceylon had well over fifty radar stations linked to seven filter rooms for the coordination of plots by the end of 1942.[25] The number soon increased to seventy with the network stretching from the Assam-Burma border to Calcutta. Forward of the Air Ministry Experimental Stations (AMES) were the Wireless Observer Units with posts across the Manipur Road and southward toward the Chin Hills as far as Arakan Hill tract. The reports flowed through a number of filter rooms: Imphal, Chittagong and Calcutta covering Bengal, Vizagapatam, Madras and Trincomalee covering the east coast, and Bombay and Cochin covering the west coast.

The expanding network of airfields came with its own attendant problem of ensuring their defence—both from air and ground. Till the fall of Burma, there existed a limited number of airfields in India and as they were far removed from enemy, the need for RAF regiment resources was never felt. But with the fall of Burma came attendant need of ensuring adequate defence of the rapidly proliferating airfields in eastern India even as there existed no RAF regiments elements in India at that time. The first element to reach India were a party of 70 or so gunners of the RAF regiment who were originally destined for Singapore but with its fall, had been diverted to India. These gunners went on to form the RAF Regiment Training Centre at Begumpet airfield near Secunderabad with a staff of seven officers and 46 NCOs and airmen.[26] The first AA flights were formed by the end of the year, with block numbers 4,351 to 4,500. The training of these units was carried out initially by Army NCOs, and the assault course at Secunderabad gained the reputation for being the most exacting in the Far East. The RAF regiment had two basic components—ground defence and air defence. For the latter, it had LAA squadrons armed with Hispano 20-mm. cannons, later augmented by the Bofors 40-mm. anti-aircraft guns.

As the anti-aircraft defences were being built up, HQ 13 AA Brigade, RA, arrived at Bombay and was retained in Southern Army as there was no brigade headquarters in western and southern India. In addition, seven British AA regiments also arrived, beefing up the overall anti-aircraft defences. To better control the growing number of AA units, HQ 2nd Indian AA Brigade was formed on 1 May 1942 at Calcutta by Brigadier H.H.M. Oliver, MC, and given the responsibility of anti-aircraft defence of Calcutta, Cuttack, Jamshedpur, Asansol, Kharagpur, Panagarh, Pandeveswar, Salbani and Kanchrapara.[27] Most of these were sites of the recently built airfields and radar (RDF) stations of the Air Force. The units of the brigade included 66th HAA Regiment, RA, 77th LAA Regiment, RA, 4th Indian HAA Regiment, 2nd Indian LAA Regiment and 6th Indian HAA Battery. The brigade formed the GOR at Tatanagar, Jamshedpur on 1 May 1942 itself. This was set up along with the No. 2 GOR at Digboi on the same date, making it a total of three GORs covering Assam and Bengal.[28] To have a better control and reporting system in place, the AA brigades were putting up GL radars, in absence of AMES, to be used as early warning radars. One such GL was established at Gobindpur Hill, on the outskirts of Calcutta.[29]

As part of the process to further streamline the anti-aircraft defences, Headquarters, Eastern Army, undertook two major steps. In August 1942, Eastern Army reformed AA and air defence branches at its HQ and combined them under Brigadier AD, making it part of the general staff.

On 24 July 1942, a Conference was held at HQ 1st Indian AA Brigade in Calcutta wherein it was decided that 2 Indian AA Brigade should establish an Eastern Army School for Mobile AA Artillery at Jamshedpur so that the Indian anti-aircraft batteries, till now trained and equipped for static role, could be trained for mobile warfare and be readied for employment in the future offensive operations.[30] No. 3 AATC had also been formed by now at Mehgaon, adding to the overall training capacity and hastening the process of new anti-aircraft regiments. The new AATC catered to Madrassi recruits only but unlike the other two AATCs, it carried out training in both heavy and light AA techniques. The first regiment to be raised at Mehgaon was the 10th Indian HAA Regiment that was formed in November 1942. The anti-aircraft cooperation to the No. 3 AATC as provided by a flight of 22 AACU at Katni but as it meant a waste of flying effort, the landing ground at Mehgaon was extended to handle Wapitis as that target towing aircraft could now operate from the local ALG and save time. Wapitis were of World War I vintage and rather slow but as this air cooperation was meant for training of recruits, they served the purpose.

In the Eastern Army there were, by now, two AA brigade HQ in Bengal and one in Assam. Keeping in mind the geographical spread of airfields and other installations in Assam, there was a need for more AA units and brigade headquarters.

Meanwhile, the 3rd Indian AA Brigade was raised in September 1942 at S & D Club, Pilot Bandar Road, Colaba.[31] With the 13 AA Brigade, RA, already in location, there was a need to rationalise the AA defences and the tasking of the brigades. The HQ Southern Army ordered two board of officers to decide on the requirement and siting of AA defences and to consider all related aspects like 'aircraft, guns, searchlight, decoys, smoke, camouflage and other PAD and ARP means' while doing so. The Commander, 3rd Indian AA Brigade, was to assess the requirement for Bombay District and Deccan (Independent)

Area while the Commander, 13th AA Brigade, RA, was to do the same for Madras district.[32] The detailed siting of the AA guns was determined by a separate board of officers who was instructed to ensure that 'each HAA gun should have two alternate and two dummy gun sites.' As the gun sites were being planned and built, the AA brigades issued orders that all isolated gun sites were to be commanded by an officer, in view of the prevailing civil disturbance. There were reports of telephone lines being cut at several places in the Madras and Bombay districts, but otherwise the civil disobedience movement did not affect the anti-aircraft defence per se.[33]

In September 1942, and subsequent to the raising of Indian AA regiments, it was decided to raise two types of workshop sections for attachment to light AA and heavy AA regiments, respectively. The light AA regiment workshop section was designed to carry out first and second echelon repairs of fire control instruments, guns and mechanical transport while the heavy AA regiment workshop was designed to carry out first and second echelon repairs of radars, guns and vehicles. The workshop sections were based on those of the Royal Electrical and Mechanical Engineers with one difference; the ordnance sections with the light AA regiments were larger to allow sub-sections to be detached, for the light AA batteries.[34] This facilitated the *in situ* repairs of radars and guns, adding much to the operational preparedness of the units.

The Japanese reacted strongly to the growth of Allied air strength in eastern India, especially when the air operations by the Allies in the Arakan challenged the Japanese command of the air. From October 1942, a series of air raids were directed against Digboi oil installations and the airfields in Assam. The air raids extended all over Bengal and Assam, up to Chittagong, with Calcutta being raided regularly.

The raids over Digboi and adjoining areas were contested by the AA batteries and several hits were claimed. On 25 October 1942, a large formation of Japanese planes was reported approaching Digboi and was engaged by the guns of 1st Indian LAA Regiment. One plane was hit with '*starboard engine in flames'* while another was *'damaged, unlikely to return to base'.* The regiment had to wait till 29 October for the confirmation of the kill when the wreckage of the plane was found.[35] The kill was credited to 1st Indian LAA Regiment—amongst the first, if not the first, kill over the homeland. The regiment had been raised at Malir Cantonment as part of 251st Indian Support Group of 31st Armoured Division in mobile LAA role, and one of its batteries (the 3rd Indian LAA Battery) had already seen action in Burma. Its two other batteries 2nd and 6th—were deployed at Upper Assam and Chittagong, respectively.[36] This kill was by the 2nd Indian LAA Battery. The regiment claimed three more aircraft in October itself, with 12 damaged.[33] Its stellar performance continued for the entire duration of the war.

The Japanese air raids, and the engagement by the RAF and the anti-aircraft guns simultaneously, meant that most of the Japanese raids were intercepted and neutralised before they caused any serious damage. But the raids continued in the following months, with formations of over 20 bombers often coming into attack an airfield. Dinjan, Chabua and Mohanbari came in for special attention with repeated attacks. The 9th AA Brigade's log for the last week of October records a raid almost every day with multiple raids on some days.[37]

25 October 1942		
1400 hours	54 bombers and 30 fighters	Dinjan and Chabua
1405 hours	30 'Zero' fighters	Chittagong
	9 fighters	Mohanbari
		Dum Dum
26 October	20 Bombers	Dinjan and Chabua
	27 fighters	Mohanbari
28 October	16 Army Type 97	Chabua
	18 Army Type 97	Dinjan
	7 Army Type 97	Dinjan station
29 October	One (Recce?)	Tamu

One of the most active AA regiments during the period was the 1st Indian LAA Regiment deployed at Chabua and Dinjan. The regiment did well in defending the airfields, scoring a number of kills and hits as recorded in its war diary.[38]

24 October 1942. Dinjan. 1945 hours. Approximately 27 Japanese bombers and 20 fighters carries out a high-level bombing and strafing attack on Dinjan airfield. Duration of the attack 30 minutes. Guns engaged—281 rounds expended—estimated effect—3 destroyed and 9 damaged.

25 October 1942. Chabua. 26 Japanese bombers and some fighters carried out high level bombing and ground strafing attack on Chabua aerodrome. Guns engaged—196 rounds fired—estimated effect—3 fighters damaged. Very few casualties. A Tp HQ burnt down to ground and all kit lost. Morale—excellent.

26 October 1942. Japanese aircraft carry out reconnaissance of Chabua and Dinjan aerodromes. At Chabua, one plane hit and damaged by LAA fire.

27 October 1942. Approx 15 bombers carry out high-level bombing of Chabua while 16 bombers raid Dinjan. As they are out of the range of LAA guns, the deployed guns are not able to engage them.

While continuing with air raids over Assam, the Japanese now turned their attention towards targets deeper in the hinterland which were now well within their range. Calcutta was too lucrative a target to be left attended for long and the Japanese first raided it in December 1942. One of the more well-known incidents, it has found mention in popular literature and culture. The damage was not severe but it did cause panic amongst the local populace, with thousands fleeing the city.[39]

[W]hat the people of the city had been fearing ever since the fall of Burma came to pass on the 20th of December. A force of eight Imperial Japanese Army Air Force (IJAAF) Ki-21 Type 97 medium bombers, code-named "Sally" by the Allies, scattered their bombs over the city. They damaged the oil plant at Budge Budge, located on the Ganges river a little south of the city, and one eyewitness reported a hole in the road opposite the Great Eastern Hotel, but the physical damage inflicted on the city was trivial compared to the devastating blow to the morale of the inhabitants: approximately one-and-a-half million people panicked and fled, among them a majority of the conservancy workers who hailed from upcountry villages. The effect on the civic services was catastrophic, and there were serious fears of an epidemic caused by the mounds of rotting garbage that accumulated. The IJAAF bombers returned on a number of occasions on the days that followed, notably on the night of the 24th December.

The AA brigade report on the first air raid is more pointed:[40]

'At 2358 on 22/12/42, stand-by for operations was given by FCR.

Raid 71

3 enemy aircraft plotted throughout at a height of 15,000 ft.

Enemy aircraft approached the mouth of R. Hooghly from the south east. They appeared to overshoot the river at first as they struck land near Birampur, turned over Contai and then following the west bank of the River northwards. They were over Falta at approx. 0037 at a height of 16,000 ft. From here they flew parallel with the K. and F. Railways. At this time, a fighter was right behind them *so the guns were not allowed to fire.*

Raid 75

Fire was opened from the Dum Dum site as planes passed over them.

The RAF report that there were three twin-engine planes in a V formation. Fighters engaged the starboard plane with cannon fire and hits were observed. The plane rocked.

30 rounds HAA were fired. No Light AA.'

As the report mentions, the AA guns were not given permission to fire during the initial raids as the fighters were in the air. It was only in some of the later raids that the anti-aircraft guns were allowed to fire—when there were no fighters in the air. Restricting AA guns to ensure safety of own aircraft was strictly followed, and barring an odd instance, worked well throughout.[41]

[P]Raid 51

Raid 51 first plotted by RAF at 0249 hours at 125,000 yds. range as hostile at 0307 hours at 100,000 yds. range.

At 0309 hours stand-by for operations given by FCR.

Three enemy aircraft came in over mouth of Hooghly orbiting near Sagar island, from which point apparently they picked up their bearings. They then followed river northwards to Diamond Harbour where they left it and headed north following the K & F Railways over Bishnupur to Alipore.

Here they lined up on 1 and 2 docks, Kiddepore, and the head of the River Hooghly at Shalimar Point.

A stick of bombs was dropped in the dock area between Garden Reach Road and Garden Reach Circular Road.

RIAF fighters were up and guns were refused permission to open up by the RAF. No guns fired. The fighters at one time were within 3 miles but failed to intercept.

A good track was obtained from the GLs throughout the time the targets traversed across the GDA.

Bombs were dropped at approximately 0328. Four fell between LAA sites 6 and 7 in Kidderpore docks.

A bomb fell beside the Kidderpore Tram Depot, several cars were damaged.

A leaflet bomb was also found near here, and was handed over to Bomb Disposal.

The most common form of attack was carried out by Army Type 97 bombers (Mitsubishi Ki 21 II or 'Sally'), flying in formation and escorted by 'Zero' fighters. The No. 978 Balloon Squadron, RAF, was located at Calcutta at the time of the raid and had deployed the balloons to protect the docks, Howrah Bridge and Writers' Building among other installations though there were no impacts with balloon cables dropped by the Japanese aircraft. The squadron suffered a few casualties during the air raid but remained operational throughout.

One of the regiments deployed in Calcutta at that time was the 7th Rajput HAA Regiment with its 18th, 19th and 20th HAA Batteries. Its war diary notes, "The regiment in action for the first time" on 21 December 1942 but as the batteries did not fire, the day was disappointing to say the least.[43] It was during the raid on 24 December that the AA defences got luckier.

Stand by for operations was given at 2100 on 24 December 1942. When two raids were plotted following the coast south-east of Sagar Island at a height of about 15,000 ft., these raids were 73 and 74. 73 following 74, each raid consisting of two planes.

R. 74 flew up the middle of Hooghly over Chakrabera and headed straight for Diamond Harbour. R. 73 was still following but slightly further to the west.

At 2134 O and N sites were given permission to engage, and all sites immediately after. R. 74 now flew over Budge Budge passing between O and N sites. The track passed to the west of Howrah, over S site and then headed north-north-east for Barrackpore and Ichapur.

B, C and E sites at Barrackpore and Ichapur fired a barrage 2144 and raid 74 turned away to the east and flew out in the direction of Tiger Point. R. 73 followed the same course and a similar barrage was fired by B, C and E at 2147. This raid also turned and flew out in the direction of Tiger Point. The RAF interceptors whilst they were going out and strikes were made.

C site at Noapara claim one destroyed during this raid as something was seen to burst into flames in the sky, fall to the ground and lie burning for a considerable time. No confirmation however has yet been obtained.

Sites stood down at 2208.

The 20th HAA Battery of the 7th Rajput HAA Regiment claimed the Japanese plane—the first such claim over hinterland. The kill, however, could not be confirmed as the wreckage was not found. With long periods of inactivity, it was never easy for the regiments to maintain a high standard of vigil—and performance. An entry in the war diary of the 7th Rajput HAA Regiment notes that the '*men often became stiff firing the guns leading to inaccurate firing*' and the spotters relied on the binoculars as the only means to search for the targets. The war diary records that the observers often failed to pick up targets at all at long ranges.[43]

One common shortcoming noted was that the newly arrived British AA regiments and the new raising of AA regiments took time to come up to the desired standards, especially when firing with the GL Mk II radars in an 'unseen' engagement; a procedure that was new to most of the regiments. In one of the regular AD exercises, commander of an Indian AA brigade noted, 'AA radars MUST make more effective efforts to pick up targets from early warning plots.'[44]

The inexperience of gunners, and the Gun Position Officers (GPOs), in particular, often resulted in AA guns not being able to react fast enough to the evasive measures taken by the raiding Japanese, giving the Japanese planes the opportunity of getting away without any damage. One such report mentions quite categorically, 'The enemy force of T97 engaged, took evasive action by changing height. Shooting poor for accuracy. GPOs shown lack of experience in dealing with this type of evasive action.'[45]

This period was also a test of patience for anti-aircraft batteries and troops as a large number of them often spent days without any enemy air activity in their area. It severely tested the command for it was a challenge to keep the men motivated and maintain the required standards of readiness. As one regiment commander noted:[46]

> The morale of the men is good although many wonder just what are they doing sitting on the gun site when there appears to be no apparent likelihood of hostilities. To these men war seems to be dragging on and with no prospect of an early finish.

The inactivity was partly due to the, by now, large presence of Allied Air Forces. With an active Allied Air Force fighter force, the first interception was generally by the fighters and restrictions were imposed on the AA regiments, prohibiting the guns to fire. Detailed orders were issued in this regards and with multiple agencies like British AA units, RAF, USAAF, US Army AA units, Indian AA units and West African AA units, it was always a challenge to coordinate their activities. One such order laying down detailed instructions for firing by the AA guns deployed at the airfields is given below. This was issued for the AA units deployed at Jorhat airfield where 51 Group of USAAC was located.[47]

Orders for firing by AA guns will be as follows:

BY DAY

HAA. Fire will be opened on any group of aircraft identified as hostile or which have committed a hostile act, or when US Ops room asks for fire. Single aircraft will not be engaged except—

(a) When they are being actively offensive;
(b) When there is a reasonable chance of bringing them down;
(c) On the request of US Ops Room for Pointer Rounds. AADC Digboi and Dinjan will then fire 4 rounds from a single selected gun.

LAA. Normal rules of engagement.

BY NIGHT

HAA. Fire will not be opened except when permission has been obtained from US Ops Room.
LAA. Fire will not be opened at night even as above unless

(a) The aircraft is making a direct attack on the gun, or
(b) Against parachute mines or flares.

Ops room has the right by day or by night, to order guns to cease fire for operational reasons.

The orders for not opening fire on a solitary plane were very explicit as it was experienced that a single plane often preceded the attacking formation to gather information about the layout of the airfield and the location of AA guns. When the attack was later delivered, the located guns received special attention, and were targeted to be put out of action when most needed. To induce the guns to disclose their positions by firing, the recce aircraft adopted the following actions[48]—

(a) flying low over the area (but usually out of range of LAA),
(b) firing cannon at no target of importance,
(c) dropping a few indiscriminate bombs.

In order to deny the enemy information of the location of guns and the layout, all AA brigades had laid down specific policies for engagement of single planes. Below is one such order, issued by 9th AA Brigade, RA.[49]

The following policy will, therefore, be adopted—

(a) A single aircraft will NOT be engaged by either heavy or light A guns, or small arms of AA units, unless it is clear that its object is to make a damaging attack on the VP or VA which the guns are sited to cover (i.e. the aircraft is not a recce aircraft)

(b) Positions of guns will NOT be disclosed when a single aircraft appears, by the running about of the personnel at gun positions, or by removal of camouflage from guns and all instruments. In the case of HAA guns, the aircraft will be followed by one instrument only (usually the TI) while the reminder of the detachments stand by with guns and instruments covered in case a genuine attack develops.

(c) LAA guns will in NO case open fire on aircraft out of range, except in the case of deliberate low flying attacks by several aircraft on the VP or the guns, when even low bursting Bofors shells may act as deterrent and assist the DCs to judge correct deflections when the aircraft comes within range.

It was not that restrictions were imposed only on AA guns as some areas were declared 'No Fly Zones' as was done in November 1942, when 5 miles around Digboi was demarcated as prohibited areas for own planes, except fighters. But then as with all such orders, the AA brigade noted the very next day, 'DC-3s are not complying with control of aircraft orders and are flying over Digboi'.[50] Such instances only made it harder for the Control and Reporting staff to exercise control over the AA guns and the aircraft.

The control and reporting system laid down the procedure for issue of air raid warning and was primarily dependant on the Air Force chain of radars and observer network for early warning. It was not always so that the Air Force network was able to pick up the Japanese air raids, more so because of the terrain in Assam. This resulted in the AA guns not getting enough reaction time to take on the intruders. The 1st Indian LAA Regiment, deployed at Tinsukhia in December 1943, faced the same problem when a Japanese raid came in at about 1150 hours. The raid was by 18 Army Type 97 bombers and 22 Army Type 97 SE Fighters coming in on a wide front. A few miles south of Tinsukhia, the formation split into two. One formation of 3 bombers and five fighters first came over Tinsukhia, and went on to the original 'breaking up point', joining the formation. The second formation of 1 VIC of nine Army 97 bombers and a second wave of five Type 97 bombers, given top cover by three formations of six Army 97 Fighters each in line astern formation, with four or five fighters below the bomber formation went for an attack on the Dinjan airfield. Two gun sites engaged the planes at extreme range and fired 5 rounds each. The neighbouring sites failed to identify the planes, which were approaching Tinsukhia out of the sun, until they had passed overhead.[51] By the time they identified and laid on them, the planes were out of range.

In a similar incident, the AA guns at Dum Duma Club in Dinjan were unable to identify the hostile planes until they were almost overhead though the detachment performed better. Two planes were seen to be hit and falling, and were claimed as Category B hits.[52] The Japanese tactics of coming out of the sun was a common practice and was often used to deny the gunners the opportunity to pick up the aircraft and also delay the identification till the very last moments.

One of the methods to compensate for inadequate warning was to use the GL sets for local warning as a properly sited radar proved very effective in detecting the intruding Japanese planes and providing warning to the AA detachments. During the raid at Dum Duma Club, too, it was one such GL that had passed the warning to the temporary gun position. Even as the planes had not been picked up visually, the warning from the radar had facilitated the engagement by the AA guns. The only factor limiting their efficacy was the duration for which they could be kept switched 'on'. Another problem was of interference between the GL Mark II sets and American Signals equipment, leading to the need to select the sites for the radars accordingly, and switching them on selectively. The AA brigades started co-ordinating the siting and roster of GL sets, networking some of them with the formation AAOR and using the GL sets for early warning. These GL sets thus supplemented the radar network of the Air Force.

Augmenting the radar network was the Indian Observer Corps initially formed to cover only Calcutta. By November 1943, the Indian Observer Corps consisted of three control units, six mobile companies in Bengal, static units in four main areas—Calcutta, Chittagong, Vizagapatam and Madras, and fourteen other mobile companies under training. In December 1943, it was decided to convert the static organisation and certain care and maintenance formations into seven additional mobile companies. By May 1944, the number of control units had increased to four, and that of operational mobile companies to seventeen with ten others under training or in process of formation.

During the battle in Arakan, four companies of the Observer Corps were involved, and redeployment of a number of posts in forward areas became necessary. Such posts, when forced to retire, succeeded in most of the cases in destroying their equipment before retreating. Similarly, when the enemy advanced towards Imphal, nearly 50 posts had to be evacuated or redeployed. The posts east and south of Imphal lost much equipment but the other posts were able to save a high percentage of equipment while all abandoned equipment was destroyed and denied to the enemy.

With the demand for more anti-aircraft regiments came the need to find the required manpower. There was already a strain on the existing pool and in order to cut back the men from select duties came the proposal of using the WAC(I)s in air defence duties. They were already serving in anti-aircraft units in Great Britain and carrying out duties that even included manning of guns. Though it seemed unlikely that women would be carrying out all such duties in India, it was felt that at least the direction finding, manning of the AA Operations Room and the plotting board were duties that could be taken on by women. The need thus was of recruiting and training women. The WAC(I) had already been formed in April 1942 and, in 1943, No. 4 AATC was formed at Calcutta to train the WAC(I) cadets in anti-aircraft duties with the 2nd Indian AA Brigade given the responsibility for the same.[53] Training of the WAC(I)s included three weeks of on-the-job training in AAORs and they were soon employed in the plotting rooms of the control centres and AA operations rooms. The AATC was also used to conduct a week-long refresher cadre for NCOs but as the planned number of WAC(I)s had been trained and there was a low demand of replacements and reinforcements, No. 4 AATC was closed down after five months in July 1943.[54] The employment of WAC(I)s was a big change—and more of a cultural challenge than anything else—but they performed their duties creditably. Except for an occasional administrative snafu, there were no problems encountered with using WAC(I)s at the AAORs. One of the WAC(I) ladies was Gloria (Paddy) Lumb (nee Stanford) who joined the service when she was just 17 and worked for two years, at the No. 3 AAOR, Digboi. Her great niece mentions 'Aunt Paddy's' experience thus:[55]

> The Japanese had already bombed Calcutta in December 1942, trying to take out the Howrah Bridge, the vital link between Howrah station and its rail links to the rest of India. Calcutta was also the site of the Dum Dum airport, with its nearby munitions factory.
>
> Aunt Paddy worked at the Gun Operation Room which was near Digboi in Assam, site of India's oil refineries. It was an underground base, well-hidden. The railway lines were also crucial.
>
> There was a real risk of malaria from mosquitoes, so they had to sleep in cloth bags and keep their arms covered at all times. It was also very cold in their underground bunker. So Aunt Paddy was very grateful when she was given a military overcoat—too large, perhaps, but at least it kept her warm.

In 1943, additional AA regiments were raised and, with the increased availability of AA guns, the desired scale of eight HAA and 12 LAA guns each to be deployed at all airfields was slowly being met. This scale was a much reduced scale of defence as compared to the anti-aircraft defence in Great Britain and Europe as the standard teaching for combined operations in UK was that an airfield should have a minimum of 16 heavy and 36 light AA guns but with the caveat that it was applicable when airfields were few in number and were located near the enemy. These conditions not being met in India, plus the presence of a large number of airfields, resulted in the reduced scale of 8 and 12. Even this was arrived at after much deliberation as the GHQ in India had earlier surmised that 'four HAA guns should provide a deterrent effect but if more than a deterrence is required, at least 16 heavy AA guns should be provided.' The 9th AA Brigade, RA, notes that 'this is apparently not known to the Americans' hence, the decision to have only eight heavy AA guns at an airfield was arrived at.[56] Though the detailed reasoning of the decision is not known, this order came to pass and was followed thereafter. The brigade took up another case—to have 20-mm. LAA guns issued to all light AA guns at the scale of one per 40-mm. gun without any increase in establishment but it was not accepted.[57] Some 20-mm. light AA guns though came to be issued to select batteries to beef up the defences, and to be used in ground role.

The anti-aircraft defences in the rear areas also kept getting upgraded as newer regiments were raised and sent to various ports and cities. One of these was the 17th Indian LAA Regiment raised in 1943 and it was deployed in Madras. One of the British NCOs, Sergeant Alf Allen, recollects:[58]

> Our first operational posting was to the city of Madras; we were stationed at Fort St. George on the coast, overlooking the Bay of Bengal. The battery headquarters was established in a group of buildings inside the ancient fort and our "Ack-Ack" detachments were manning gun sites in remote positions around the port of Madras and on the embattlements of Fort St. George.
>
> Apart from my involvement in the ongoing training of our Indian troops, such as gunnery courses, etc., I was also part of the site "Ack-Ack" Operations Room team, where we maintained a twenty-four-hour vigil, plotting the aircraft movements over our sector. We were also in direct contact by field telephone with all of our gun sites, which were manned by Indian gunners, whose use of the English language was somewhat limited.

By now, the anti-aircraft defences were fine-tuning their drills and the war dairies of the period are replete with references to 'need to pass crisp, short orders as too much time is wasted from the time orders are given to fire and the guns actually fire.'

Another note records that there is 'too much talking in gun pits' and it disturbs the readiness of the gun site.[59]

The arrival of the 9th AA Brigade, RA, in Assam did much to streamline the anti-aircraft defences. It had first moved from Barrackpore, on the outskirts of Calcutta, to Jorhat where it established the Headquarters in the American Baptist Church. The organisation of AA Defences in Assam was based on geographical, and operational distribution of responsibility with the entire are being divided into two[60]—

(i) Assam Area—to include late Assam district as forward as possible so as to include Road-Head (RH), RH base and depots on any Line of Communications (L of C) of an operational front
(ii) Forward area—All forward area excluding (i)

The 9th AA Brigade, RA, was made responsible for AA defence in Assam Area, under orders of 4 Corps and for local administration, the anti-aircraft units in Assam were placed under command of Assam Division. While the units in Digboi, Dinjan, Tinsukhia were administered through the North Assam Brigade, the units in Dimapur were under HQ Base Sub Area, Dimapur.

By June 1943, the 9th AA Brigade was protecting key supply points and had deployed at fifteen airfields with the desired density of AA defences. The guns had destroyed 19 hostile aircraft by now and the combined effort of AA regiments and Allied fighter aircraft resulted in the Japanese suffering an attrition rate of up to 20 per cent of their attack force. Such a high loss rate forced the Japanese to cut back its air raids and adopt low-flying tactics.[61]

The 13th AA Brigade was deployed to protect the airfield supporting the 15th Corps and was spread till Chittagong. All the airfields were protected by eight heavy and 12 LAA guns except Chittagong which had 24 3.7-inch heavy AA and 32 40-mm. LAA guns defending the airfield and the supply base given its criticality for the ongoing operations. With the first

Arakan campaign launched in December 1942, the airfields came under heavy raids by the Japanese Air Forces, with the pattern of attack following the standard Japanese tactic of enemy bomber formations coming at 27,000 feet with low-level sweeps by the Zero fighters. The regiments of the 13th AA Brigade did well to protect the airfields, shooting down 13 Japanese aircraft in May 1943 alone.[62]

The air raids by the Japanese were being met with increasing effectiveness by the Allied air forces and the anti-aircraft artillery but concerns remained about the threat posed by them, especially amongst the legislators and the civil population at large. The question of Japanese air raids was often raised in the Legislative Assembly and the member sought to be assured that all measures were being taken to protect the civil population. Tours were organised for the Defence Consultative Committee of the Legislative Assembly of India to visit the anti-aircraft units and establishments. In addition, exhibitions were organised all over the country to showcase the anti-aircraft radars and guns. The Japanese planes shot down were always a special attraction, drawing huge crowds.

There were occasional incidents that kept the civil population, and the air defences, on the edge. It often led to some unintended actions, too, like the one in Madras as reported by a local newspaper on 8 November 1943:[63]

> On Rasapaa Chetti Street in George Town, V. Kannan, aged 28, a resident of Triplicane, refused to run to a safety shelter despite repeated warnings. The Court ordered him to pay a fine of Rs. 50 or undergo rigorous imprisonment for a month for what it considered irresponsible behaviour on his part.

Such precautions were not wholly wrong as the Japanese planes tried to raid Madras again on 12 November in one of their last attempts to cause any damage. The raid was repulsed by the alert anti-aircraft gunners and no bombs were dropped on the city. The situation was slowly but surely becoming stable in the 'homeland' as anti-aircraft defences became more and more effective, with almost all areas now well-defended and an efficient

Plate 1: Madrassi LAA gunners

Plate 2: LAA battery of 25 Infantry Division at Maungdaw, September 1945

Plate 3: 19 HAA Battery of 7 Rajput HAA Regiment in Assam, February 1945

Plate 4: Madrassi AA gunners defending the No. 1 Squadron, RIAF airfield, 1944

Plate 5: Shot down Japanese bombers on public display in Calcutta, January 1943

Plate 6: Commander-in-Chief visits LAA site Calcutta, November 1944

Plate 7: Commander-in-Chief visits Maratha LAA Battery at Kohat, December 1944

Plate 8: Maharaja of Faridkot visits LAA site Karachi, January 1944

Plate 9: No. 3 AATC, Mehgaon, 1943

Plate 10: 16 HAA Battery/6 Indian HAA Regiment, RIA, 1944

Plate 11: LAA Training of recruits at FATC, Mathura, 1943

Plate 12: 15 Indian LAA Regiment at Malir Cantt, October 1944

Plate 13: AA Defences, Bombay Port, 1944

Plate 14: LAA Battery undergoing Combined Operations Training at Combined Training Centre, Bombay, April 1944

Plate 15: 19 HAA Battery of 7 Rajput HAA Regiment in Arakan, December 1944

Plate 16: 6 Punjab LAA Regiment Contingent at the Delhi Victory Parade, March 1946

system in place. The military situation, on the other hand, was not so assuring. The first Arakan offensive had failed and 15th Corps was back to positions near Cox's Bazar by May 1943, from where it had started. The Japanese air raids were, however, going down in their intensity and the homeland was much safer though the real battle was yet to come.

Three days later on 11 November 1943, the Japanese raided the airfields in Assam. Amongst them was Kumbhigram which had a battery of the 8th Sikh LAA Regiment defending it. As the Japanese carried out high-level bombing, the LAA guns were not effective and the Japanese planes were able to bomb the installations at the airfield. One of the bombs fell near the gun position and hit the ammunition stacked near the gun position, setting fire to the ammunition and stores kept there. At great risk to their personal safety, Gunners Mehnga Singh and Hari Singh of the Bofors detachment managed to extricate the stores before they caught fire. They were both awarded the IDSM.[64]

Notwithstanding an occasional success to the Japanese air forces, the anti-aircraft defences were by now quite capable of defending the homeland and it was now time to extend the cover provided by the anti-aircraft defences to the field armies as they prepared to return to Burma and defeat the Japanese but, before that, the anti-aircraft had to prepare for the offensive.

NOTES

1. As recounted by Maj C.H.T. MacFetridge, *Oral History* at Imperial War Museum.
2. Ibid.
3. File No. 601/2290/WD, War Diary of 1st Indian HAA Regiment.
4. As recounted by Maj C.H.T. MacFetridge, *Oral History.*
5. File No. 601/2290/WD.
6. Bisheshwar Prasad, *Expansion of the Armed Forces and Defence Organisation, 1939-45,* pp. 311-13 and Memorandum on '*Expansion of Indian Artillery 1940-42*' by Royal Artillery Directorate, GHQ, New Delhi, accessed at History Division, Ministry of Defence, Government of India.
7. *Statistical Review of Personnel,* Army of India, vol. I, 1 July 1939

to 1 October 1941 and Memorandum on '*Expansion of Indian Artillery 1940-42*'.

8. N.W. Routledge, *History of the Royal Regiment of Artillery: Anti-Aircraft Artillery*, Brassys, London, 1998, p. 236.
9. *Statistical Review of Personnel*, Army of India, vol. I, 1 July 1939 to 1 October 1941.
10. By 28 March 1942, there was only one working radar unit in India, No. 258 AMES based at Mathurapura, South of Calcutta. (Peter Preston-Hough, *Commanding Far Eastern Skies*, Helion and Company, 2019, pp. 34-5.
11. *Statistical Review of Personnel*, Army of India, vol. II.
12. Routledge, op. cit., pp. 236-7.
13. Ibid.
14. Prasad, op. cit., p. 96.
15. John Clancy, *The Most Dangerous Moment of the War: Japan's Attack on the Indian Ocean, 1942*, Casemate Publishers, Philadelphia, 2015 (Kindle Edition).
16. Ibid.
17. Ibid.
18. 'And then, Madras was Bomber', *The Hindu*, 5 October 2012, accessed at https://www.thehindu.com/news/cities/chennai/and-then-madras-was-bombed/article3965756.ece on 10 July 2019.
19. Yasmin Khan, '*India at War: The Subcontinent and the Second World War*', Oxford University Press, New York, 2015, pp. 171-4.
20. 'And then, Madras was Bomber', *The Hindu*, 5 October 2012.
21. Routledge, op. cit., pp. 236-7.
22. File No. 601/2099/WD, War Diary of Headquarters 2 Indian AA Brigade.
23. Ibid.
24. Memorandum on 'Expansion of Indian Artillery 1939-1941' and its Addendum 'Expansion of Indian Artillery 1942-43', prepared by Royal Artillery Directorate, GHQ, Delhi, History Division, Ministry of Defence, Government of India, New Delhi, held by History Division, Ministry of Defence, Government of India, New Delhi.
25. Preston-Hough, op. cit., pp. 36-8.
26. Nigel N.M. Warwick, *Constant Vigil: The RAF Regiment in the Burma Campaign*, Pen and Sword Aviation, Barnsley, 2007 (Kindle Edition).
27. File No. 601/2099/WD, War Diary of Headquarters 2 Indian AA Brigade.

28. Ibid.
29. Ibid.
30. File No. 601/2099/WD.
31. File No. 601/2100/WD, War Diary of Headquarters 3 Indian AA Brigade.
32. Ibid.
33. Ibid.
34. Bisheshwar Prasad, *Technical Services, Official History of the Indian Armed Forces, 1939-45,* Combined Inter-Services Historical Section (India and Pakistan), New Delhi, 1957.
35. File No. 601/2436/WD, War Diary of 1st Indian LAA Regiment (Now 25 LAA Regiment).
36. Ibid.
37. File No. 601/2093/WD, War Diary of Headquarters 9th AA Brigade, RA.
38. File No. 601/2436/WD.
39. David Lockwood, *Calcutta Under Fire: The World War Two Years,* Rupa Publications, New Delhi, 2019.
40. File No. 601/2098/WD, War Diary of Headquarters 1st Indian AA Brigade.
41. Ibid.
42. File No. 601/2302/WD, War Diary of 7th (Rajput) HAA Regiment.
43. Ibid.
44. File No. 601/2099/WD.
45. Ibid.
46. File No. 601/2439/WD, War Diary of 4th Indian LAA Regiment (Now 29 LAA Regiment).
47. File No. 601/2093/WD.
48. Ibid.
49. Ibid.
50. Ibid.
51. File No. 601/2436/WD.
52. Ibid.
53. File No. 601/2099/WD.
54. *Statistical Review of Personnel,* vol. III.
55. Margaret Deefholts, *Women of Anglo-India: Tales and Memoirs,* CTR Inc Publishing, New York, 2010, pp. 95-6.
56. File No. 601/2093/WD.
57. Ibid.

58. Mike Allen, 'My Family History,' accessed at http://www.mike-allen.org.uk/alf/start.htm on 21 July 2019.
59. File No. 601/2093/WD.
60. Ibid.
61. Ibid.
62. File No. 601/2094/WD, War Diary of Headquarters 13th AA Brigade, RA.
63. 'And then, Madras was Bombed,' *The Hindu*, 5 October 2012.
64. *The London Gazette*, Supplement 36426, p. 1,264, 14 March 1944.

CHAPTER 6

Prelude to the Return to Burma

The defeat in Burma and the retreat from Burma in 1942 had brought to fore various limitations in the training and organisation of the Commonwealth formations and units. The troops were not trained for jungle warfare and the units and formations were ill-equipped and ill-organised. There was an over-reliance on mechanical transport and the weapons were more suited to the warfare in the plains than in the jungles. One of the first steps taken, post the retreat from Burma, was to reorganise the infantry divisions. First, the number of mechanical transport was reduced in three of the Indian divisions—the 7th, 20th and 23rd, and were organised with a mix of animal and mechanical transport. They were classified as 'Mixed Transport Divisions', later to be named as the Animal and Mechanised Transport (A & MT) Divisions. In addition, the 17th and 39th Indian divisions were converted to light divisions of two brigades, with six mule companies and four jeep companies each, in order to be able to operate away from the road. Other changes in the light divisions included abolishing the anti-aircraft and gun platoons in the British units. The gun platoon was replaced by a medium machine-gun platoon.[1]

Still other changes included the formation of a jungle field regiment in each division of which two batteries were equipped with eight jeep-towed 3.7-inch howitzers (modified with pneumatic-tyre wheels for towing) while the third battery had 16 3-in. mortars to give artillery support where 25 pdr. field artillery guns could not be used.

REORGANISATION

In August 1943, the South East Asia Command (SEAC) was formed under Vice-Admiral Lord Mountbatten with him as the

Supreme Allied Commander. The command embraced the new Fourteenth Army, composed of British and Indian divisions and the Fifth and Sixth Chinese Armies under General Stilwell, the Deputy Supreme Allied Commander and commander of all US forces in South-East Asia. General Stilwell remained Chief of Staff to Chiang Kai Shek. The SEAC had all land and sea forces allotted to it by the Combined Chiefs of Staff for the defeat of Japan. Its immediate strategy was planned on three fronts: in the north-east, General Stilwell was to make a long advance from the Ledo Road to Myitkina and the upper Irrawaddy, in the centre, Fourteenth Army was to launch an offensive from Imphal, across the Chindwin towards Mandalay, and in the west, Fourteenth Army was to produce another offensive in the Arakan, to clear the Mayu peninsula and open a sea route for supplies through the Maungdaw. The last part was to include sea-borne landings but the removal of the majority of landing crafts from SEAC made it an impossibility.[2]

In 1943, the Fourteenth Army under General Slim consisted of two active Corps, the 4th and the 15th with the 33rd undergoing collective training. To ensure the required air support for all its operations, a network of airfields had come up in Assam and Bengal. The unified air command of SEAC had 33 tactical support squadrons, six of light bombers, 10 of heavy bombers and a group of transport squadrons, primarily tasked with supplying China. The airfields were defended by a mix of British and Indian AA regiments, and the 14th West African AA Brigade with three regiments.[3]

One innovation in the employment of LAA that was adopted in the Fourteenth Army in 1943 was the formation of composite LAA/A Tk (Anti-Tank) regiments having two LAA and two Anti Tk batteries each. Four British LAA/A Tk regiments were formed in 1942-3; 24th, 55th, 56th and 82nd. These were followed by the 100th and 122nd in 1944.[4] The reorganised LAA/A Tk regiments were intended to help divisions distribute sub-units with balanced firepower and mutual support in close jungle fighting where battle groups often had to move independently to occupy isolated defensive localities, the 'boxes'. The employment of these regiments varied as per the formation, for example, 15th Corps policy had both the LAA and A Tk batteries being used in anti-tank or ambush role with no troops dedicated for the LAA tasks.

This experiment did not last long as the LAA batteries were removed in 1944, and the regiments reverted to their original role of A Tk regiments. All other LAA regiments in the Fourteenth Army were fully mobile and had been boosted in strength to three batteries of 18 guns each. Some regiments were equipped with 18 guns on self-propelled made instead of towed mountings, to be employed as one complete battery or in separate troop configuration of three guns each.

The reorganisation of units as Light AA/A Tk regiments was carried out for selected Indian anti-tank regiments as well. A total of four Indian anti-tank regiments, viz., 1st, 2nd, 7th and 15th Anti-Tank Regiments were identified to be converted as LAA/A Tk regiments and underwent conversion training to a varying degree but, except for 2nd Anti-Tank Regiment, none of the regiments were employed in the new role.[5]

The 1st AA/A Tk Regiment was first raised at Hyderabad as 'K' Anti-Tank Regiment, IA, on 1 August 1940 and was renamed as the 1st Anti-Tank Regiment on 2 April 1941. On 2 August 1943, the regiment was transferred to Eastern Army and moved from Sialkot for Chas, in Bihar, where it arrived by 29 August 1943 coming under command of the 26th Indian Infantry Division. Whilst this move was taking place, on 15 August, the regiment was notified that it was earmarked for conversion to an AA/A Tk regiment. On 12 September 1943, the regiment moved to Barambe, near Ranchi, for conversion to the new role and organisation but it was a slow process. The regiment handed over eleven 6-pounder guns to the 24th Anti-Tank Regiment, RA, in September 1943. A total of six officers and 21 IOR underwent an anti-aircraft course but the regiment did not receive any LAA guns till December 1943. Even as the conversion training was being carried out, the Indian divisional anti-tank regiments were once again the subject of another reorganisation from August 1944, when they were reorganised to form three batteries, each equipped with twelve 6-pounder anti-tank guns and twelve 3-inch mortars.[6] By October 1944, the regiment had converted to this new organisation.

The 2nd AA/A Tk Regiment was first raised at Hyderabad as the 'L' Anti-Tank Regiment on 1 April 1941 and was renamed as the 2nd Anti-Tank Regiment on 2 April 1941. The regiment with the 5th, 6th, 7th and 8th Anti-Tank Batteries had served in Burma in 1942 as part of the 17th Indian Infantry Division. By early 1943, the regiment had become divisional troops to the 23rd Indian Infantry Division. The regiment was notified in July 1943 that the 5th and 8th Batteries were to convert to the light anti-aircraft in the near future, equipped first with 40-mm. Bofors guns and eventually 20-mm. Hispano guns for both anti-aircraft and ground role. Though both 5th and 8th Batteries had been earmarked for the conversion, only the 8th Battery was converted in March 1944, receiving twelve Bofors guns, and renumbered as 208 LAA Battery. With this, the regiment was also renamed and given the designation of 2nd Indian LAA/A Tk Regiment with effect from 1 April 1944. The regiment served with 23rd Infantry Division during the period and took part in the Battle of Kohima where the 208th LAA Battery was deployed at Langgol with one troop at Palel. This was the only instance a LAA battery of Indian LAA/A Tk Regiment taking part in operations.[7]

The 7th AA/A Tk Regiment served with the 25th Indian Infantry Division up until March 1944 when it moved to Madras where by 1 April 1944 it was under conversion as an AA/A Tk Regiment but in August 1944 it was reorganised back as an anti-tank regiment.[8] By the time it left Madras in late 1944 for the Arakan to rejoin the 25th Indian Infantry Division, it had the 9th, 25th and 26th Anti-Tank Batteries, equipped with both 6-pounder anti-tank guns and 3-inch mortars. The 15th (Punjab) Anti-Tank Regiment was raised on 1 August 1942, by the conversion of the Machine Gun Battalion, 15th Punjab Regiment. It was part of the General Reserve of the GHQ, India, until 19 July 1943 when it came under the command of 4th Corps. In September 1943, it moved to Piska Road near Ranchi coming under 151 Lines of Communication Sub-Area for administration. At this time, the regiment was organised as a four battery regiment, composed of the 28th, 29th, 30th and 31st anti-tank Batteries. The regiment shed the 28th, 29th and 30th Anti-Tank Batteries and converted to LAA/A Tk with 203 & 204 LAA Batteries along side the 31st Anti-Tank Battery. It came under the command of 33rd Corps and was located at Imphal when it was demobilised in November 1944.[9]

The other changes included the formation of a jungle field regiment in each division of which two batteries were equipped with eight jeep-towed 3.7-inch howitzers (modified with pneumatic-tyre wheels for towing) while the third battery had sixteen 3-inch mortars to give artillery support where 25 pounder field artillery guns could not be used. For anti-aircraft artillery, another change was to switch to a 54-gun establishment for select Indian LAA regiments from the 36-gun establishment and the forming of composite AA batteries with four heavy and eight light AA guns.[10] The first of such batteries was the 102 Indian Composite Battery that was sent to Addu Atoll in the latter half of the year.[11]

CONSOLIDATION

In July 1943, Headquarters 14th West African AA Brigade with three HAA regiments arrived in India and was deployed for the defence of airfields in the Surma valley. Unlike the British and Indian AA brigades, it did not have any British of Indian AA regiments. The United States Anti-Aircraft Artillery (AAA) battalions had also started arriving in India, to be deployed at the airfields used by the US Army Air Force. Even not considering the West African and the US AAA units, India now had the second most dense anti-aircraft defences, second only to the home defences of Great Britain.[12]

By September 1943, the Japanese air threat had considerably reduced and, accordingly, expansion in Indian anti-aircraft artillery was halted. With this, the scale of anti-aircraft defences was reduced and some reduction, both in British and Indian AA, began to be affected. The first to be demobilised was the 7th Sikh LAA Regiment which moved to FATC, Mathura, with the regiment-less 22 LAA Battery being demobilised on 15th August 1943.[13] This was followed by two more AA batteries in the same month and the 14th Indian LAA Regiment by the end of the year.[14] The pre-war class units were selected for the first demobilisation in order to make reinforcements available for other arms. The disbandments brought in associated problems. As only the pre-war class units were affected by the initial cuts, the absorption of higher ranks of Punjab classes rendered surplus on disbandment was becoming difficult as while a number of lower ranks could be retained in reinforcement batteries for maintenance, there were few options for placing and employing the NCOs. A small number were placed in other arms and services but no satisfactory solution to this problem was found.[15]

Inadequate number of signallers was a problem faced by all branches, not only the anti-aircraft artillery, throughout the war, and the severe shortage of signallers continued right till the end of the war. The main problem was, however, the provision of, an adequate number of VCOs and NCOs of Madrassi descent. As per some reports, provision by direct enlistment during 1943 was not an unqualified success but the provision of NCOs from unit personnel, through special courses, was producing a gradually mounting number of capable Madrassi NCOs and, as these men became available, the corresponding number of British NCOs were released.[16]

A major change was the conversion of certain Indian HAA regiments to a mobile role, mobile HAA having been hitherto confined to British regiments only. In addition, it was decided to give greater mobility to the remaining Indian AA units, other than those rendered surplus to operational requirements, by the addition of an increased scale of transport. The units thus reorganised were to be known as 'Semi Mobile' units and a suitable War Establishment was devised for them. The units thus converted were brigaded under Mobile and Semi Mobile AA Brigade Headquarters, respectively, and the training commitments and responsibilities were undertaken by the Brigade Commanders concerned.[17]

The last AA units to be raised at AATC, Malir, was the 15th Indian HAA Regiment which was formed in October 1943. A few Indian HAA regiments were raised at Mehgaon in the same period but were soonafter converted to LAA regiments.[18] The personal diary of Major Sidney Ernest Wright, one of the regimental officers, gives a detailed account of the training regime in the last days at AATC, Malir:[19]

> From now onwards, the days were occupied in training—technical, and all the other branches. By January, 42 and 43 Batteries were ready for practice camp. Each unit spent a week at the AA School in Karachi, firing each day. Very good results were obtained, and right sleeves were shot down. Satisfactory reports were issued by the School to GHQ on the unit's progress. The shooting was certainly of a higher standard than one would expect from a young Indian regiment. It was more consistent and accurate than the 8th Belfast when I saw them fire at practice camp. It now remains to see what will be their reaction to action against enemy targets. After this camp, the batteries return to Malir and concentrate on the weaknesses which were found in the firing. After a further month of training they return to AA School, and take the final practice camp before mobilization of the regiment.

For the British officers, Malir was the favoured training centre, preferred over Deolali and Mehgaon. For officers sent for new raisings, it was a pleasant change. The accommodation was far better than what a regimental officer, or a gunner, was used to. The AATC was housed in a huge camp that had been built to accommodate an armoured division. With the threat of the Germans breaking through from Russia into the Middle East, and thence through Khyber into India vanishing, the camp had been taken over as an AA training centre, and by the Americans as a reception camp for troops. The officers lived in bungalows with electric lights and fans, furnished by the MES; and the offices and lecture rooms were all modern and adequate. There was a garrison cinema in the lines, and a swimming pool in the American section which could be used by the officers posted at the AATC. Such facilities made the stay worthwhile, the weather notwithstanding.

TRAINING CENTRES

With the cutbacks, the training centres were also reformed. As the expansion plan had neared completion, it was necessary to plan for the time when the role of the AA training centres would be limited to the training of the reinforcements for active units. No. 2 AATC at Deolali which catered to the pre-war class personnel was already doing so but Nos. 1 and 3 AATC—for Madrassis—were till now providing trained recruits in bulk for new units. Of the two AATCs catering for the Madrassi units, the start was made with No. 1 AATC at Malir and it was disbanded in April 1944.

By June, the last of the newly raised Madrassi units had made their demands on No. 3 AATC for basically trained recruits. Accordingly, in June 1944, No. 2 AATC was disbanded and No. 3 AATC reorganised so as to include both pre-war classes and Madrassis. The two wings of the AATC now provided basic military training as well as corps training for all recruits for Indian AA artillery units. The new organisation took over the existing accommodation at Mehgaon and from June 1944 onwards, the AATC trained recruits and provided reinforcements for all Indian HAA and LAA regiments.

WAIT FOR THE OFFENSIVE

The period 1942-3, was one of setbacks and frustration for the Army in India. The only offensive operation carried out was the advance in Arakan with the limited goal of retaking the Akyab Island. The island lay at the end of the Mayu Peninsula, marked by a narrow but precipitous and jungle-covered range of hills, the Mayu Range, and held a port and all-weather airfield. This was to play a central role in the plans for retaking Burma as air operations from Akyab could cover most of Central Burma, and medium bombers could even reach as far as Rangoon.[20]

In December 1942, the Army had taken the initiative with 15th Corps launching an offensive to retake Akyab with the 14th Indian Division, starting the advance south from Cox's Bazar on 17 December 1942. After six months of frustrating effort,

XV Corps, beaten back by a resolute Japanese Army, fell back to Cox's Bazar on 11 May 1943. The only other offensive operation were undertaken by the Chindits but they were supported by 69th LAA Regiment, RA, in their second expedition, with no Indian AA regiment employed in the operations.[21]

Away from the operations, Indian AA regiments faced their own challenges. Long periods of inactivity with no Japanese raids to counter made it difficult for the regimental and battery commanders to always maintain a high standard of vigil and discipline. Training was the only way out—to keep the men busy and keep honing the skills for the day they all may be required to take on a hostile raid. This was also becoming a challenge with long periods at same location, carrying out laying practice on an AACU aircraft as the only activity and in case it was cancelled, all that the units could do was vent their frustration in their war diaries.[22]

> Air cooperation arranged from 1000 hours is cancelled. This is unfortunate as the practice gained from a low-level attack is considerable and is not obtained from an aircraft flying normally. It is hoped that the headquarters takes note and this does not become a regular feature. . . .

The frustration that was seeping in is summed up in the remarks of a regimental commander in the following:[23]

> The men badly want to go into action. 2 ½ years of training—for static LAA manning with only one raid (and that too at a high level) amounts to training and nothing more in the eyes of many—is rather long. The counter to this is the news of going mobile. The danger that I may have to watch is any feeling that the regiments' further training may take so much time that even the Jap war may be over by the time we get a chance to fire a shot.

Thankfully, the formation of SEAC and the coming months were to see the Indian anti-aircraft artillery start replacing British regiments and be employed in a major way in all operations.

NOTES

1. Tim Moreman, *The Jungle, Japanese and the British Commonwealth Armies at War, 1941-45*, Frank Cass, 2005.
2. N.W. Routledge, *History of the Royal Regiment of Artillery: Anti-Aircraft Artillery*, Brasseys (UK), London, 1998, p. 238.
3. Ibid.
4. Memorandum on 'Expansion of Indian Artillery: Addendum to Cover Period 1942 to August 1943' and Routledge, op. cit., p. 239.
5. Woodburn Kirby, *The War Against Japan*, vol. II, H.M. Stationery Office, London, 1958, p. 242.
6. War diary 1st Anti-Tank Regiment, I.A., WO 172/2403.
7. War diary 2nd Anti-Tank Regiment, I.A., WO 172/2404.
8. War diary 7th Anti-Tank Regiment, I.A., WO /4758.
9. War diary 15th Anti-Tank Regiment, I.A., WO 172/2407.
10. Memorandum on 'Expansion of Indian Artillery: Addendum to Cover Period 1942 to August 1943'.
11. File No. 601/2510/WD, War Diary of 102 Indian LAA Battery (Later 102 Composite AA Battery).
12. Routledge, op. cit., pp. 85-6.
13. *Statistical Review of Personnel*, Indian Army, vol. III, 1 January 1944 to 30 September 1945, compiled by Army Statistical Organisation (A.S.O./2), published by Manager Government of India Press, New Delhi, 1945.
14. Ibid.
15. Bisheshwar Prasad, *Expansion of the Armed Forces and Defence Organisation,* Royal Artillery Directorate, GHQ, New Delhi, pp. 98-9.
16. Memorandum on 'Expansion of Indian Artillery 1939-1941' and its Addendum 'Expansion of Indian Artillery 1942-43', prepared by Royal Artillery Directorate, GHQ, Delhi, History Division, Ministry of Defence, Government of India, New Delhi, held by History Division, Ministry of Defence, Government of India, New Delhi.
17. Ibid.
18. 15th and 16th Indian HAA Regiments were converted as LAA Regiments by 1944 (*Statistical Review of Personnel*, vol. III).
19. *Personal Diary of Ernest Wright*, accessed at https://www.lennonwylie.co.uk/ WW2DiaryWright1.htm on 21 July 2019.
20. Routledge, op. cit., pp. 239-41.
21. Ibid.
22. File No. 601/2438/WD.
23. Ibid.

CHAPTER 7

Back to Burma

The setbacks in 1942-43 brought in a period of despondency, with the morale in the Army badly affected by the inability to garner any success against the Japanese. Even the attempt to notch up a victory in the attempt to retake Akyab with the 15th Corps putting in all its weight behind it had failed. The setbacks had given rise to the myth of the unbeatable Japanese soldiers, helped to a large extent by the rumours spread amongst the local populace, and the jawans.

One of the first tasks before the Allies was to restore the confidence of the soldiers and break the myth of the 'invincible' Japanese. The Chindits fit in perfectly in this role. Creation of British Army Brigadier Orde Charles Wingate, the Chindits were formed for raiding operations against the Imperial Japanese Army, especially long-range penetration: attacking Japanese troops, facilities and lines of communication, deep behind Japanese lines. The first Chindit unit, the 77th Indian Infantry Brigade, was formed in the area around Jhansi in the summer of 1942. Abandoning the standard battalion structure, it was formed into eight columns, each based on an infantry rifle company with a support group with the heavy weapons, a reconnaissance platoon and a sabotage group. The support group had two LAA guns for anti-aircraft role.[1]

Just before launch of the first operation, one of the columns was broken up to make up the remaining seven up to full strength and Operation Longcloth, the first Chindit expedition, was launched on 8 February 1943 with 3,000 men.[2] The columns operated behind enemy lines for three months, carrying out various acts of sabotage and disruption; mainly of the north-south railway line. The operations across the Irrawaddy river

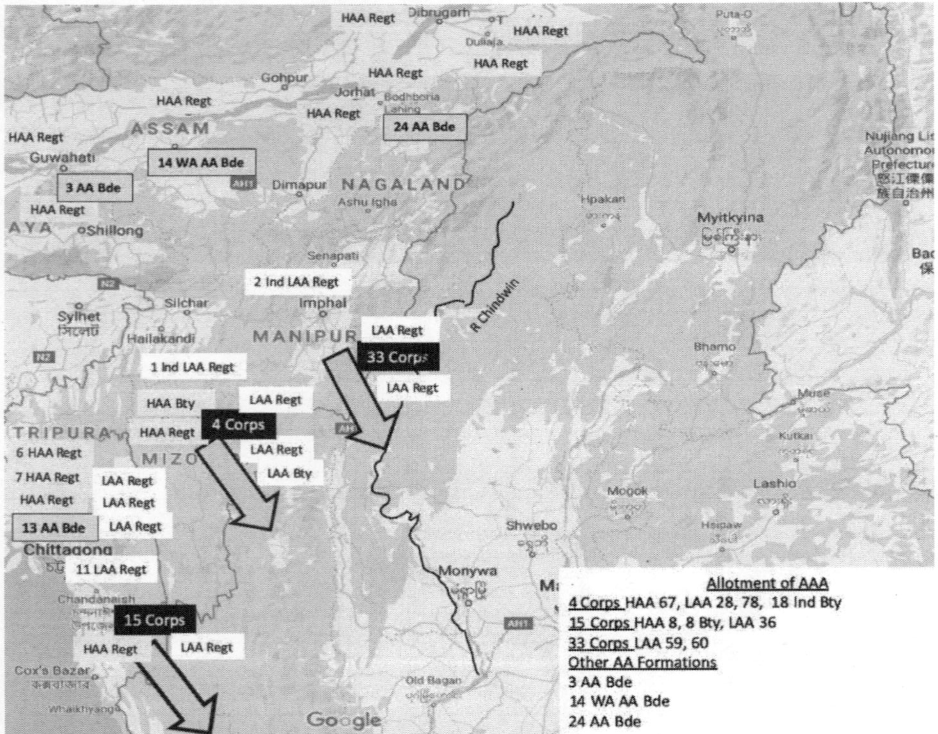

Map 4: Allotment of AA Resources for 14th Army Offensive 1943-4.
Source: Author's collection.

faced greater Japanese opposition, as a result of which the force was ordered to withdraw after about three months. While the military effectiveness of the operation was debatable at best, it did have a positive effect on the morale of the Allied troops.

As the members of the first expedition were making their way back to India, a second long range penetration unit, the 111th Indian Infantry Brigade was already being formed, under Brigadier Joe Lentaigne. Originally, the Chindits were to be of two-brigade strength but, using his clout, Wingate planned to raise the force to be of six brigades for which three (14th, 16th and 23rd Infantry Brigades) were taken from the British 70th Division by breaking it up and one (3rd West African Infantry Brigade) from the 81st West Africa Division.[3] Officially called the 3rd Indian Division, the Chindits were to wait for some time before they were to be launched.

In August 1943, the South East Asia Command (SEAC), was formed under Vice-Admiral Lord Mountbatten as Supreme Allied Commander. The command embraced the new Fourteenth Army, composed of British and Indian divisions and Fifth and Sixth Chinese Armies under General Stilwell, the Deputy Supreme Allied Commander and commander of all US forces in South East Asia. General Stilwell remained Chief of Staff to Chiang Kai Shek. The SEAC had all land and sea forces allotted to it by the Combined Chiefs of Staff for the defeat of Japan. Its immediate strategy was planned on three fronts: in the north-east, General Stilwell was to make a long advance from the Ledo Road to Myitkina and the upper Irrawaddy, in the centre, Fourteenth Army was to launch an offensive from Imphal, across the Chindwin towards Mandalay and in the west, Fourteenth Army was to produce another offensive in the Arakan, to clear the Mayu peninsula and open a sea route for supplies through the Maungdaw. The last part was to include sea-borne landings but the removal of the majority of landing crafts from SEAC made it an impossibility.[4]

In 1943, the Fourteenth Army under General Slim consisted of two active Corps, the 4th and the 15th, with the 33rd undergoing collective training. To ensure the required air support for all its operations, a network of airfields had come up in Assam and Bengal. The unified air command of SEAC had 33 tactical support squadrons, six of light bombers, 10 of heavy bombers and a group of transport squadrons, primarily tasked with supplying China. The airfields were defended by a mix of British and Indian AA regiments, and the 14th West African AA Brigade with three regiments that had arrived in India in July 1943.

The one innovation in the employment of LAA adopted in the Fourteenth Army had been the formation of composite LAA/A Tk regiments but this experiment did not last long and the regiments reverted back to their original role of A Tk regiments, shedding the LAA batteries. The other change in the Fourteenth Army, of having fully mobile LAA regiments with three batteries of 18 guns each, continued. Some of these regiments were equipped with self-propelled guns instead of the towed versions that made them more suitable for mobile operations with the field Army.

In December 1943, Slim's Fourteenth Army took the offensive with 15 Corps tasked to advance from its holding positions south of Chittagong and clear the Mayu Peninsula and retake Akyab in the south. The corps had the 5th and 7th Indian Divisions and the 81st West African Division. The AA order of battle, included British AA and AA/A Tk regiments—one heavy and one light AA Regiment each as corps troops and one AA/A Tk each with the two Indian divisions. The AA/A Tk regiment of the West African Division was still under training and was not available for operations.[5] There was no Indian AA regiment with the corps.

The aim was to move down the coast of Burma, take the island of Akyab and prepare to continue the offensive, supporting the overall campaign to retake Burma.

The topography of the Arakan presented many problems as had been evident in the first, unsuccessful offensive. The steep-sided central mountain ridge ran all the way down the peninsula, with the only decent road being along the western side and their being dense or broken jungle on the eastern side. Only three places offered lateral access between the two sides; a rough and winding track starting from Bawli and crossing the Goppe

Pass, a steep and narrow track over the ridge at Ngakyedauk about half-way along the peninsula, and a well-constructed hard road running from Razabil through Letwedet to Buthidaung, making use of tunnels cut through the ridge. With such terrain constraints, there were not many options in making the Corps plan. The 5th Indian Division was tasked to take the western route, with the armour and a larger share of artillery, and the 7th Indian Division was to cross the Goppe pass and move along the eastern route. The 81st West African Division, taking a longer route, was to enter the Kaladan valley to outflank the Mayu.

The offensive began with the 7th Indian Division attacking on the eastern side of the peninsula as planned and the 5th Indian Division attacking in the west to secure the port of Maungdaw. The initial objective was the Japanese fortifications guarding the Maungdaw-Buthidaung road, crossing the ridge along the spine of the peninsula. As the Japanese were expected to react strongly to the offensive, the 26th British and 36th Indian Divisions were warned for a possible move to the Arakans.

The offensive of the 15th Corps was halted by the Japanese 55th Division which was dug in along a mountain spur extending west to the sea near Maungdaw, blocking the land route to Akyab. For nearly two months, the 15th Corps tried unsuccessfully to break the Japanese defences which had by now been reinforced by the 54th Division.

The AA and AA/A Tk regiments found themselves not only taking on the Japanese air attacks but also getting involved in the ground battle, with the batteries taking on the role of direct firing artillery and also moving stores over the Ngskyedauk pass.[6] As the 15th Corps was switching over its main attack on to the eastern flank, the Japanese 55th Division launched the Ha-Go counter-offensive, directed against the 7th Indian Division. The Indian division had taken up position in four brigade boxes, separated by the jungle and scrubs. The Administrative Box was at the rear with its dumps of ammunition, supplies and stores.

The 55th Division counter-attack had, besides the element fixing the 5th and 7th Indian Divisions, a second smaller force going against the 81st West African Division and a third force infiltrating, attacking the rear areas, to cut off the supplies and destroy the divisions as they retreated. It was planned to be a repeat of the Japanese strategy used earlier with success but this time around, the situation was different as General Slim refused to authorise a withdrawal and the formations of 15 Corps dug in for a long haul. Resupply for the troops boxed in was carried out by air and the units held on with the AA batteries getting enmeshed in local ground battles. In the Battle of the Admin Box itself, one-and-a-half LAA Battery with one HAA troop was used to great effect in taking on and repelling repeated Japanese attacks. The AA gunners contributed as equally as the infantry in successfully defending the Admin Box.[7]

Having held back the Japanese, Slim moved up the 26th Indian and 36th British Divisions to restore the situation. It was the Japanese who were now encircled. Having to switch their forces to meet the new front, the Japanese offensive collapsed.

The 5th Indian Division resumed its assault on Razabil, with the 7th Indian division advancing to take the Letweder-Buthidaung road. In March 1944, as the 15th Corps prepared to move ahead, the monsoon broke, bringing the advance to a standstill.

The successful defence against the Japanese advance had an effect on the anti-aircraft defences as well. The year 1944 was the start of the Allies increasingly gaining air superiority and, as a result, a decision was taken to reduce the overall anti-aircraft defences in India. All British static AA regiments in the India/SEAC theatre, except one, were planned to be disbanded by May 1945. A number of HAA regiments of Royal Artillery were to be converted to Medium regiments and the HQ 9th AA Brigade, RA, with 53 HAA Regiment, RA, was moved to Gulunche near Poona in January 1944. The brigade headquarters later moved to Ahmednagar and received two more British HAA regiments—all three were to be converted by year-end.[8]

While the Battle of the Admin Box was going on, Fergusson's 16th Brigade left Ledo for Burma on 5 February 1944. The rest of the brigades were brought in by air to create fortified bases with airstrips. This was the start of the second Chindit expedition, codenamed Operation Thursday. As a change from the previous operation, the columns were to operate from strong 'keeps' or 'forts', which were to be supplied by air transport operations, rather than air drop as in the first operation. For this, the use of landing strips within the fort perimeter was to be made and it was decided to give the landing strips some degree of anti-aircraft defence. Four troops of three officers and 74 men each were formed for this purpose by 69th LAA Regiment, RA. The four, 'W', 'X', 'Y' and 'Z' Troops had six 40-mm. guns each. The first troop to be flown in was the 'W' which was air transported to 'Broadway' on 6-8 March. The guns had barely been deployed that the Japanese air raids came in. The six guns were not enough to cover the entire area and the high-level attacks meant that the Japanese planes were beyond their reach, but five planes coming in at low levels were destroyed.[9]

As 'Aberdeen' was ready two weeks later, the 'X', 'Y' and 'Z' Troops moved in on 24-26 March. While the main AA guns were the air portable 40-mm. Bofors, with two-wheel mounting, the 'X' Troop was equipped with 10 × 0.5 inch machine guns. The LAA troops performed well, claiming six Category I kills at 'White City' and nine at 'Aberdeen'. With Myitkina airfield secured by the Chinese divisions, the Chindits linked up Stilwell's forces and held on for 78 days against heavy shelling, air and ground attacks as the Japanese tried to dislodge the Allies, but failed. As the 36th Division arrived from the North, Upper Burma was secured. Their task complete, the Special LAA troops flew back to Assam.[10]

The main battle in the theatre was, however, at Imphal. The stage had been set for the battle in the dying days of the Battle of the Admin Box. Slim had foreseen that the Japanese, stung by the failure in the Arakans, would attack from the Chindwin valley to open a thrust into India. Accordingly, Imphal was chosen to form the main defence against the expected Japanese offensive, and act as the springboard for a subsequent advance.

Imphal, the capital of Manipur state, lies on a plateau surrounded by hills. It was the main British base in the area. Road access from the hinterland was from Dimapur and Kohima. From Imphal, tracks and roads led from it to Tiddim and the Manipur valley in the south, to Palel and Tamu in the east and to Ukhrul in the north-east. Imphal had been built up as a major logistic base, held by 4 Corps with two airfields and a number of landing grounds. Two divisions of 4th Corps were placed forward; the 17th Indian Division at Tiddim and the 20th at Palel-Tamu. One Brigade of 23rd Indian Division, the Corps reserve, was at Ukhrul. The remainder of the 23rd Indian Division was held back, with the 254th Indian Tank Brigade. The AA support for the 4th Corps included three regiments directly under the corps and a regiment each with the three divisions; the 2nd Indian LAA Regiment with the 23rd Indian Division being the only Indian AA regiment.[11] The 4th Corps was deployed in a series of well separated 'boxes' on high ground, with screens on the main approaches. The LAA guns were split between the screen positions and the gun areas, and the HAA guns were at the main airfields at Imphal, Palel and Bishenpur.

In the months preceding the Japanese offensive, the Japanese Air Force was quite active, with formations of Army Type 97 bombers, escorted by fighters, carrying out regular raids. While they were more often than not intercepted successfully by the Allied Air Forces and engaged by the ground-based air defences, the fighter sweeps by low-flying 'Zeros' did cause some damage as they came along the valley floors, making it difficult for the guns to engage them.

The Japanese offensive started on 6 March with the 15th and 33rd Japanese Divisions moving against Imphal and the 31st Japanese Division towards Kohima. The 17th Indian Division withdrew in face of strong assaults by the Japanese; in a move that lasted three weeks. It was joined by elements of the 23rd Indian Division who helped clear the route to Imphal and by early April the 17th Indian Division had taken up positions at Imphal. Meanwhile, the 20th Indian Division, facing the 31st Japanese Division, had also withdrawn to Imphal, in boxes at Shenam and Bishenpur. These were particularly targeted by the Japanese Air Force, though over 10 aircraft were shot down by the LAA.[12]

Along the Kohima axis, the 31st Japanese Division cut off the road between Imphal and Kohima, developing the attack on the main defences, held weakly by the Assam Regiment and some of the paramilitary Assam Rifles. The 161st Indian Brigade was originally posted at Kohima, but it was moved out of Kohima to Dimapur shortly before the attack. On 3 April, the 31st Division attacked Kohima Ridge which overlooked the main supply route between Imphal and Kohima. The siege lasted for weeks till Kohima was relieved by 14 May and the Imphal Road was cleared by the 2nd Division. Having blunted the Japanese offensive, the 4th Corps now got down to pushing back the enemy. In the drive against Tiddim, the 19/7 Rajput HAA Battery, a part of the 23rd Indian Division, was employed in a direct support role.[13]

The Japanese troops were driven out of the Kohima region by the end of May and they fell back southward. The 33rd Corps was then joined by the 4th Corps and attacked the Japanese troops further along the Dimapur-Imphal road on 22 June near Milestone 109. When the battle at Milestone 109 was won by the British and Indian troops, the siege of Imphal was lifted. The Japanese 33rd Division continued to exert pressure on the Indian troops and nearly broke through the defences at Bishenpur, but in the end the Japanese army suffered such a high casualty that the offensive was called off on 3 July 1944. The remnants of the Japanese 33rd Division were withdrawn back to Burma.

As the ground offensive went on, a major air battle was fought over the Imphal plains against Japanese bombers and fighters operating from their advance bases in the Irrawaddy valley. The Japanese Air Forces' plans to cause any serious damage were effectively neutralised by the combined efforts of Allied Air Forces and the anti-aircraft artillery.

With the Japanese attempt to reach India halted at the Battle of Kohima, the air threat also abated.

A period of consolidation of AA defences took place. It meant a pruning of anti-aircraft units and it had an adverse effect on the AA Order of Battle in India as a number of British AA regiments were broken up to release manpower for infantry. These included four HAA regiments of Royal Artillery. The remaining HAA regiments in Burma retained their AA role but were organised to act as medium artillery. Four such regiments acquired sections of 7.2-inch howitzers.[14] This change resulted in Indian AA artillery having to give more regiments to the 9th and 13th AA Brigades, RA, to fill up the voids created by the removal of British regiments. By December 1944, these two brigades were made up entirely of Indian AA regiments.[15]

With the Japanese still reeling from their failure at Kohima, Slim now moved fast to clear the hills and valleys beyond Kohima and push back the Japanese back across the Chindwin. The 33rd Corps was tasked for this, with the 4th Corps withdrawn from the sector. The Corps had two AA regiments of the Royal Artillery, besides the AA regiments with the divisions. As part of the readjustments in the rear areas, 13th AA Brigade, RA, extended its area well into the Arakans, up to Maungdaw, in order to cover the supply routes and airstrips of the 15th Corps. The brigade had as many as five regiments and two batteries of Indian AA Artillery on its orbat and was deployed in East Bengal, as far forward as Chittagong.[16]

The 14th West African AA Brigade also moved forward with its regiments, including the 2nd, 4th and 8th Indian LAA Regiments, coming under the command of 14 Army, and deployed its regiments at Dimapur, Imphal and Palel.[17]

As the plans of the 14th Army was starting to push back the Japanese, the air situation improved as the number for forward airfields usable by the Japanese came down. Allied Air Forces now had air superiority over Assam, severely limiting the possibility of major Japanese air raids. As a result, the requirement of anti-aircraft defences came down. The scale of anti-aircraft defences was accordingly reviewed and in August 1944, the state of readiness was brought down and norms for manning of guns relaxed. The number of radars to be kept on 'ready' mode were cutback and only one radar at each VA was to be kept 'on'—to be capable of use with detachments immediately. The scale of LAA guns was reduced by half at each site.[18]

THE PUSH TO RANGOON

The offensive by 14th Army to clear Burma lasted nearly a year, till the capture of Rangoon in May 1945. The air superiority gained by the Allied Air Forces resulted in a fall in AA actions but the

AA regiments with the 14th Army remained busy—as they were increasingly employed in ground role.

In November 1944, the Allied Land Forces South East Asia (ALFSEA) was formed by grouping together the 14th Army, Line of Communication areas and the Northern Combat Area Command. One of the first decisions of the ALFSEA was to review the AA plans for the future, and a joint Air/AA Defence Committee was formed for the same under the chairmanship of BRA, ALFSEA. It recommended that the AA defences be removed from Assam, Surma valley and eastern Bengal so as to move two AA brigades and the LAA regiments forward to defend the rear echelons of the Fourteenth Army's advance. The LAA regiments with the divisions were also removed, and placed under Army/Corps Headquarters. The 2nd Indian LAA Regiment thus became one of the four AA regiments given to 33rd Corps during its advance across the Chindwin. To cover the void thus created in the AA defence of Assam and Burma, the 3rd Indian AA Brigade and the 24th AA Brigade, RA, from Ceylon were moved up to Burma.[19]

By the end of the year, LAA guns were withdrawn from Dinjan, Jorhat, Chabua, Devgaon and Moran airfields and all HAA guns were withdrawn from Dinjan, Jorhat, Chabua, Devgaon, Mohanbari and Moran in first week of January 1945.[20] There was almost a complete roll back of anti-aircraft defences in Assam and Bengal, with the focus now on supporting the operations of the 14th Army and protecting the forward airfields.

The operations after the stand at Kohima saw the Fourteenth Army employing 33rd Corps for clearing the Manipur heights of the Japanese and establishing bases on the Chindwin. The 5th Indian Division had taken the Tiddim-Tamu-Kalemyo route with four AA regiments, two of which were AA/A Tk, that were broken up at a later stage. The division was met by the 11th East African Division at Kalemyo; having advanced east from Imphal to Tamu. As Tamu was cleared, the 2nd Indian HAA Regiment was moved forward to Tamu from Dimapur and deployed there. The 11th East African Division moved down the Kabaw valley to link up with the 5th Indian Division. The 11th East African Division then moved up to Kalewa, the ferry centre that was to be a crossing site for 33rd Corps. One battery of the 3rd Indian LAA Regiment was deployed for the bridgehead protection at Kalewa.[21] The other crossing site was at Kalemyo, and both bridgeheads were to be protected by a HAA regiment each, supplemented by a LAA regiment.

Meanwhile, the 19th Division of 4th Corps had advanced from Imphal on a second thrust eastwards to cross the Chindwin at Paungybin and reach the Myitkina-Mandalay railway line, linking up with Stilwell's Northern Combat Area Command.

In December 1944, the 33rd Corps launched its main drive from the bridgehead at Kalewa on two axes; the 2nd Division taking the route to Mandalay while the 20th Division went for Monywa. Both the divisions had intimate AA support for defence of route protection, cover for airstrips and gun positions, and the bridgehead was defended by the 52nd Heavy AA Regiment and the 2nd Indian LAA Regiment.[22] Having captured Monywa, the 20th Indian Division turned south-east along the Irrawaddy, towards Mandalay. To provide AA support for the next phase of operations, the two AA regiments with the formations were centralised and the two deployed on the bridgehead, including the 2nd Indian LAA Regiment, brought forward to now provide route protection.

In January 1945, the 3rd Indian AA Brigade had AA troops at Digboi and Tinsukhia, with only two Indian AA regiments forming part of it—the 5th Indian HAA Regiment and the 5th Indian LAA Regiment. The brigade had already shed 28 static 3.7-in. HAA guns that were being returned as surplus to the depots and had four 3.7-in. guns at Sookerating as brigade reserve. With 14th Army needing greater AA resources, the 5th Indian LAA Regiment moved up for defence of the forward airfield supporting it. For this task, it was given mobile LAA guns by the 69th LAA Regiment, RA, at Manipur Road as the regiment would have to keep moving up, keeping pace with the advance of the 14th Army.[23] The other Indian AA regiments in Central Burma at this time were the 5th Indian HAA and the 1st and 2nd LAA Regiments.

As part of the 14th Army, the 4th Corps offensive was led by the newly arrived 19th Indian Division, but the formation was soon transferred to 33rd Corps and the 4th Corps was switched to the right flank of the 14th Army, advancing down the Gangaw valley west of the Chindwin. In late February, the 7th Indian Infantry Division established bridgeheads over the Irrawaddy. The 3rd Indian LAA Battery, defending the bridgehead, beat back repeated Japanese air raids and effectively neutralised the Japanese air in its effort to disrupt the crossing, shooting down an aircraft on 19 February.[24] As the division passed through the bridgehead, the Battery continued to stay at the same location, ceasing to be under the command of 7th Infantry Division.

BATTLE OF MANDALAY

As the 33rd Corps closed in to Mandalay, the 19th Division reached the outskirts of Mandalay on 7 March, making it a total of three divisions that were set to assault the Japanese citadel. As the formations were set to take Mandalay, General Slim had opened up a new front with the 4th Corps launching an offensive from lower Manipur to the Irrawaddy near Pakokku, crossing over and then advancing east to take Meiktila. This offensive was carried out by the 7th Indian Division with two AA regiments supporting it and two, the 1st and 2nd Indian LAA Regiments defending the rear areas and the route.[25] On its approach to the Irrawaddy, the formation captured the airfield at Sinthe and established the bridgehead at Pakokku for the 17th Indian Division to pass through. Having crossed the river, the 17th Indian Division headed out fast to reach Meiktila on 20 February and captured it by 24 February. The 2nd Indian LAA Regiment was deployed at Meiktila, to be used in ground role. In the north, Mandalay, meanwhile, had fallen to 33 Corps on 20 February. Having secured central Burma, Slim now turned his focus on to the capture of Rangoon.

Before the next phase of the 33rd Corps plan was put into action, the 15th Corps was already advancing in the south as part of the SEAC's plans, down the Mayu peninsula and the Kaladan valley in the Arakan. The airfields supporting the advance and its supply routes were the responsibility of the 13th AA Brigade, RA, that was by now made up entirely of Indian AA regiments. Its AA regiments included the 6th and 25th Indian HAA and the 3rd, 4th, 11th and 15th Indian LAA Regiments. Akyab fell to the advancing divisions of the Corps in January 1945. To provide AA defences to the airfield and the landing strip, the 11th Indian LAA Regiment was moved to Akyab, undertaking a landing in an amphibious operation.[26]

Two swift operations followed the capture of Akyab. The first was to take Myebon and Kangaw to trap the Japanese troops between the two towns and also the 82nd Japanese Division that was moving south. The second operation was to capture Ramree Island with the help of the 26th Indian Division.

The airfield on the Island was extended into two 2,000 yds. all-weather runways with perimeter taxiways and loading bays. With a constant flow of supplies and Dakotas and bombers flying in and out, it became one of the busiest airbases in Burma. A battery of the 11th Indian LAA Regiment was deployed for its LAA defence but it soon became apparent that its importance notwithstanding, there was little or no possibility of low-level attack by Japanese fighter-bombers. The LAA battery was withdrawn and concentrated for training. In a departure from its LAA role, the battery trained on MRLS (Multi-Rocket Launching Systems) code-named 'Land Mattress' meant for use in a ground-to-ground role. The personal diary of one of the British gunners with the battery notes:[27]

> The Land Mattress was an adaptation of the 'Z Battery' used against aircraft. This system was mobile and was designed as a 'ground-to-ground' barrage system. At this stage it was considered that the war against Japan would continue for at least another year, so this was our current situation. The unit started training and sifting out unsuitable personnel.

After the capture of Ramree, the 15th Corps carried out further landings on the mainland in an attempt to trap the withdrawing Japanese but did not succeed in its aim as the Japanese moved further into the interior. The 15th Corps was broken up after this attempt, releasing units for other fronts, and was left with only the 26th Division that carried out a sea-borne invasion for the capture of Rangoon in late April. As the stage was now being set for the capture of Rangoon, changes in the AA layout were afoot in the rear areas. In March 1945, AA deployment at Dimapur base and Imphal were cancelled by HQ ALFSEA in light of the much reduced and marginalised Japanese capability to strike these from the air. This ended the task of the 3rd Indian and 14th West African AA Brigades in Assam. The scale of AA defences was also reduced at the forward areas keeping in mind the much reduced capability of the Japanese to mount any air raid against targets in Assam.

The 9th AA Brigade, RA, that had been moved to Bombay, was given a training role for Indian AA Regiments for Operation Zipper, the projected invasion of Malaya.[28] Meanwhile, the 13th AA Brigade moved out of East Bengal and took over Akyab and Ramree from 15th Corps. With the 9th and 13th AA Brigades, RA, moving out to their new locations, there was a need to bring in a new AA Brigade HQ and, accordingly, HQ 24th AA Brigade, RA, moved to Jorhat from Colombo and took under command the 6th Punjab LAA Regiment.[29] It was to wait for some time before it was to move and take over the responsibility of the AA defence of Meiktila.

The ALFSEA policy of moving out the British AA regiments had resulted in four Indian AA regiments moving in to fill the void created by the exit of the Royal Artillery regiments from Burma. The Indian AA regiments were now increasingly being employed with the field formation, and not merely in defending the rear areas. The 8th and 9th Indian LAA Regiments were moved to Bombay, to train them for a new role—of providing anti-aircraft defence in amphibious operations.

FINAL PUSH TO RANGOON

The final push to Rangoon was led by the 4th and 33rd Corps with the 4th moving on the eastern route with the 5th and 17th Infantry Divisions, the 254th Tank Brigade and 33rd Corps on the western route and the 2nd, 7th and 20th Infantry Divisions with the 255th Tank Brigade along the Irrawaddy.

During the push in the east, with the 5th and 17th Indian Divisions leapfrogging, the control over AA batteries was decentralised to keep pace with the leading elements. After a stiff battle at Pyabwe, the 5th Indian Division took Toungoo and soon reached Payagyi, just about 100 km from Rangoon. Till now, the AA batteries were with the infantry columns but as the AA defence of the airfields at Pegu and Payagyi was critical at this stage, the AA batteries were taken off from direct support to the formations and were deployed for the protection of the airfields. To reorganise the AA defences, the 24th AA Brigade, RA, took over the responsibility of AA defence of Meiktila airfield that was used for flying in reinforcements and supplies. Its units included the 5th Indian HAA and 2nd Indian LAA Regiments.[30]

On the western route, the important objectives were, first, the oilfields at Yenangyaung, and then the airfield at Magwe. The Corps took the oilfields and Magwe on 19 April, turning the area around the airfield into a virtual fortress with AA regiments providing not only AA defence but also taking on ground defence tasks. In the drive to Prome, the formation used the AA regiments in infantry role; using them for clearing ambushes, carrying out flank patrols and encounter actions.

The day the 255th Tank Brigade reached Prome, Rangoon had fallen.[31] The Japanese had given up Rangoon without a fight as the 26th Indian Division reached the city. The division and other forces had sailed in six convoys from Akyab and Ramree Island between 27 April and 30 April, landing on 2 May after Elephant Pass had been secured and the passage up the river had been cleared. The Indian troops began occupying Rangoon the same day with elements moving north to link up with the 14th Army. The responsibility of AA defence of Rangoon was also taken over by 24th AA Brigade with the 1st, 2nd and 5th Indian LAA Regiments coming under command of the brigade on 6 May 1945 with the added responsibility of AA defences of Myingyan, Meiktila (three air strips), Toungoo, Pegu, Mingaladon airfields.[32] As SEAC intelligence appreciation was that there was a reduced threat of Japanese air raids during the monsoons, the state of readiness and percentage of guns manned was accordingly reduced, except in Rangoon where the anti-aircraft defences were maintained at high alert.

On 10 May, the 2nd Indian LAA Regiment reached Lewe as it moved towards Rangoon and after a halt at Pegu, reached Rangoon on 14 May 1945; becoming the first AA Regiment to do so.[33] It deployed three troops for the defence of Rangoon with a troop defending 14th Army HQ and two troops at the jetties at Dala and the anchorage on Rangoon River at Syrian. The 24th AA Brigade, RA, was now given the responsibility of Rangoon as well, with the 52nd HAA and 2nd Indian LAA Regiments and a battery of the 1st Indian LAA Regiment coming under its command. There were 24 × 3.7-inch HAA and 78 × 40-mm. LAA guns deployed for the air defence of the docks, the airfield and the oil installations. A number of 40 mm. LAA guns were found amongst the debris left by the Japanese. They had been disabled by the Indian and Burmese gunners when they had withdrawn from Rangoon in 1942, but the Japanese had restored them to working order. Also found were four 3.7-inch and two naval 4-inch AA guns.[34]

The fall of Rangoon brought an end to just one phase of the operations as the Japanese forces still remained in Burma and the 17th Indian Division moved eastwards, across the Sittang, to clear south-east Burma. It was only by June that a semblance of control was assumed over Burma and much-awaited celebrations for the victory could be held. On 15 June 1945, a victory parade was held at Rangoon that was reviewed by Lord Mountbatten. Two SP guns of 7/2 Battery, one SP gun of 8/2 Battery and one SP gun of the 9/2 Indian LAA Regiment took part in the parade, with the 6/2 Indian LAA Battery providing LAA defence of the parade area. The contingents of the Regiment were commanded by Capt. Noddings and Lt Lashkar Singh.[35]

In June 1945, the HQ 3rd Indian AA Brigade took over Toungoo, Mingaladon, Zayatkwin and Rangoon from 24th AA Brigade and with it the command of four AA regiments including the 5th Indian HAA and the 2nd and the 5th Indian LAA Regiments.[36] The AA batteries remained deployed on AA role even though the possibility of a Japanese air attack was remote.

In end-June, the 1st Indian LAA Regiment was reorganised as a self-propelled (SP) LAA regiment on a two troops per battery basis. The SP guns and transport were taken over by the regiment from 28th LAA Regiment, RA, that was to be disbanded. The regiment also exchanged six officers, due for repatriation to the UK shortly, with officers from 28 LAA Regiment, RA, and concentrated at Myingyin for collective training.[37]

The situation remained fluid in July with anti-aircraft batteries deployed both on AA role and in ground role, in support of infantry formations. The location state of the 3rd Indian AA Brigade on 27 July 1945 gives out the spread of the AA batteries at the end of the war:[38]

Location	*HAA*		*Rdr*	*LAA*	
	Battery	*No. of Guns*	*GL*	*LAA Battery*	*No. of Guns*
Rangoon					
Docks and Anchorage	155/52 HAA	7	1	Two Tps 7/2 Indian (Mob)	12
	159/52 HAA	8	1	Two Tps 8/2 Indian LAA and two guns of 7/2 Indian LAA	14
	271/52 HAA	8	2	9/2 Indian (Mobile)	18
	13/5 Indian	8	2		
Starch Factory Power Station	-	-	-	One Tp 7/2 Indian LAA less two guns	4
Airfields					
Mingaladon	12/5 Indian	8	1	32/5 Indian	18
Toungoo				One Tp 18/5 Indian One Tp 31/5 Indian	12

- 14/5 Indian HAA Regiment less one Tp (4 mobile guns) in support of 19 Infantry Division at Toungoo in ground role

- One Tp 14/5 Indian HAA Regiment (4 mobile guns) under command 17 Infantry Division at in ground role
- One Tp 8/2 Indian LAA Regiment concentrated at Dala ready to move and deploy at Syriam oil tanks
- 28 LAA Regiment, RA, concentrated at Insein without equipment
- Two Tp 18/5 Indian LAA Regiment (12 guns) concentrated at Kalewa
- Two Tp 31/5 Indian LAA Regiment (12 guns) concentrated for training

(*Location state of 3 Indian AA Brigade on 27 July 1945*)

Source: File no. 601/2100/WD, War Diary of Headquarters 3 Indian AA Brigade for July 1945 held by History Division, Ministry of Defence, New Delhi.

The deployment for AA tasks, and training for proposed future operations continued so till August when Japan surrendered unconditionally on 15 August. The end of the war in Burma would normally have been a matter of great joy but as the prospect of going home became brighter, fear and apprehension about the future also started growing. The British AA units had already started being repatriated and this led to rumours of a cutback of Indian anti-aircraft units. Of greater concern for the gunners was discharge and loss of livelihood. The Monthly Intelligence Report No. 10 dated 11 July 1945 of 2 Indian LAA Regiment mentions:[39]

Though murmurs of discontent about being unemployed are heard, the morale is still good. I attribute this chiefly to my being able to keep them interested; for I have more than normal equipment, e.g., transport, R/T sets and so on and I am training them in the more interesting jobs coming. The morale would drop steadily.

It has got to the ears of the IORs that if they return to India they may lose their *batta*. So far that has not caused alarming concern except to those who look well ahead, but full realisation will bring considerable discontent. Prices at home are not what they were, and *batta* made all the difference to the family income.

One battery reports, 'It is thought that the IORs would be happy if more frequent visits could be made to their women folks in the villages by the Sevadarnies. They are doing a good job in keeping up the spirits of the people at home which is always reflected in the morale of the men.'

The successful culmination of operations in Burma were followed by the Allied victory over Japan in August 1945. The dropping of the atomic bombs on Hiroshima and Nagasaki, followed by the unconditional surrender was the real end of war. As the war diary of 1st Indian LAA Regiment notes:[40]

15 August 1945. News received that Japanese have accepted the Allies' surrender terms. Two days holiday granted to all ranks.

Victory parades were held at a number of places in Burma and India with the AA Regiments participating in the same. At Myingyan, the parade was held on 20 August with 33 SP 40-mm. guns of the 1st Indian LAA Regiment taking part, and a salute of 36 rounds was fired by 12 guns of the regiment during the parade.[41]

The successful campaign in Burma enabled an almost complete reduction in the AA defences of India to be undertaken. At this time, a policy was initiated to convert certain Indian HAA regiments to a mobile role, mobile HAA having hitherto been confined to British units. The number of mobile Indian LAA units was further increased. In addition, it was decided to give greater mobility to the remaining Indian AA units, other than those rendered surplus to operational requirements, by the addition of an increased scale of transport. The units thus reorganised were to be known as 'Semi Mobile' units and a suitable War Establishment was devised for them. The units thus converted were brigaded under Mobile and Semi Mobile AA Brigade Headquarters, respectively, and the training commitments and responsibilities were undertaken by the brigade commanders concerned.

These changes were brought in to give more responsibilities to Indian Anti-Aircraft Regiments in future but the anti-aircraft defence role in Burma ended in September 1945 and with it the last of British regiments were withdrawn. The 3rd Indian AA Brigade and its units found themselves engaged in 'anti-dacoit' operations.[42]

The war may have ended but with it came the fears of an uncertain future. Training programmes were changed for all regiments though it was not clear what lay in the future. Speculations about the future were rife, and due to lack of any clarity, led to many rumours. The regiments tried to cope with the change in the best manner possible, carrying out routine activities and conducting lessons. 1st Indian LAA Regiment noted:[43]

22-31 August 45. Training programmed re-adjusted to suit post war requirements as far as possible, with emphasis on education.

31 August 45. Much speculation regarding the future of the regiment.

The remaining AA tasks in East Bengal and Assam had ended by now. The 13th and 24th AA Brigades, RA, were disbanded. The 9th AA Brigade, RA, already converted to its new role of 60 AGRA (AA), was no longer needed as the ceasefire had by now extended to Malaya. The 14th WA AA Brigade returned to West Africa and by October 1945, all remaining British regiments were on their way back.[44]

The war was over in Burma. It was only the Indian AA brigades and regiments that remained. On 13 October 1945, the regiment of Indian Artillery became the 'Royal Regiment of Indian Artillery'. As a war diary notes, '*the honour being graciously bestowed on the Regiment by The King Emperor in recognition of their services during the war*'.

A new chapter was now to be written in the history of the Royal Regiment of Indian Artillery.

NOTES

1. N.W. Routledge, *History of the Royal Regiment of Artillery: Anti-Aircraft Artillery*, Brassey's, London, 1998, pp. 239-4.
2. Bisheshwar Prasad, *The Reconquest of Burma*, vol. 1: *June 1942-June 1944*, Pentagon Press, New Delhi, 2014.
3. Ibid.
4. Routledge, op. cit., p. 238.
5. Ibid., pp. 243-5.
6. Ibid.
7. Ibid.
8. File No. 601/2093/WD.

9. Routledge, op. cit., pp. 243-5.
10. Ibid.
11. File No. 601/2437/WD.
12. Routledge, op. cit., pp. 245-54.
13. File No. 601/2302/WD.
14. File No. 601/2093/WD.
15. Routledge, op. cit., pp. 245-54.
16. Ibid.
17. Ibid.
18. File No. 601/2094/WD.
19. Routledge, op. cit., pp. 245-54.
20. File No. 601/2094/WD.
21. File No. 601/2438/WD.
22. File No. 601/2437/WD.
23. File No. 601/2441/WD.
24. File No. 601/2497/WD.
25. File No. 601/2436/WD and File No. 601/2437/WD.
26. File No. 601/2447/WD.
27. John Absolon, 'Duck Boy', *WW2 People's War* at BBC, London, 2004 accessed at https://www.bbc.co.uk/history/ww2peopleswar/stories/83/a2381483.shtml on 10 July 2019.
28. File No. 601/2093/WD.
29. File No. 601/2096/WD.
30. Ibid.
31. Routledge, op. cit., pp. 245-54.
32. Ibid.
33. File No. 601/2437/WD.
34. Routledge, op. cit., p. 245-54.
35. File No. 601/2437/WD.
36. File No. 601/2100/WD.
37. File No. 601/2436/WD.
38. Location state of 3 Indian AA Brigade on 27 July 1945 (File No. 601/2100/WD).
39. File No. 601/2437/WD.
40. File No. 601/2436/WD.
41. Ibid.
42. File No. 601/2100/WD.
43. File No. 601/2436/WD.
44. Routledge, op. cit., pp. 245-54.

CHAPTER 8

Other Theatres and Lesser Campaigns

IRAQ

The involvement of Anti-Aircraft Artillery in Iraq began during the brief clash between the British and the insurgent groups in 1941 when the dissident groups laid siege to the RAF base at Habbaniya. The British relief columns were targeted by the Iraqi rebels and subjected to a few air raids, necessitating the allotment of AA defences but GHQ Middle East had no AA units to spare as all its resources were tied down in the Western Desert, Syria and Crete. It was not until a battery of 57th LAA Regiment, RA, came to Habbaniya that the RAF base had some AA defences but it was a shortlived deployment as the battery was soon recalled for operations in Syria. However, the rebels had been put down and no air threat manifested for quite some time.[1]

A more serious threat to oil fields in Iraq and Persia was visualized from the German advance, via., the Caucasus and to safeguard the imperial interests, 4 AA Brigade was reformed in Persia and Iraq Command (PAIFORCE) in August 1942 as part of the Tenth Army.

At first it had an HAA and three LAA regiments but two of the LAA regiments were sent forward as divisional troops leaving only two regiments for the defence of the oil fields. However, there was no serious air threat to Iraq after the German defeat at the Battle of Stalingrad with the Luftwaffe not in a position to attack Iraq or Persia. Consequently, the AA defences were run down from April 1943.

In April 1943, the 8th AA Brigade, RA, was given two training tasks in addition to its operational commitments. The first was to train four Polish AA regiments for which the manpower was to come from the many thousands of prisoners of war captured in 1939 and now released by Russia.

The second task was to form an HAA cadre that was to be the basis of a new Indian AA regiments—the 25 Indian HAA Regiment. It was to be a mixed British-Madrassi regiment with the draft of Madrassi recruits for the Regiment to be sent from India.[2]

The nucleus of the regiment was formed with personnel from the 258 HAA Battery of 83 HAA Regiment, RA. The first batch of Madrassi recruits was trained and ready by May 1943, forming the first battery of the 25 Indian HAA Regiment which was sent to Habbaniya as part of the AA defences of the RAF bases. The raising of the regiment was completed by March 1944 and it formed part of 4th AA Brigade, RA.

The 25 Indian HAA Regiment consisted of Regimental HQ, 72nd, 73rd and 74th Indian HAA batteries, Indian Artillery, with each battery having two four-gun Troops. To meet the urgent requirement of trained and experienced manpower, one Subedar and three Jemadars were war posted from Queen Victoria's Own (QVO) Madras Sappers and Miners present in Iraq.

A training cadre of 12 Officers and 59 BOR was sent on 4 March 1943 from Basra to India to collect IOR, reaching Bombay on 15 March. The training cadre was attached to No. 1 AATC, Malir Cantt, Karachi on 20 March and it took over 521 IOR the next day, having been sent from No. 3 AATC, Mehgaon. The training of the IOR started the next day itself, and they were joined by 67 followers from FATC, Mathura, and 53 IOR from No. 3 AATC, Mehgaon. The training on 3.7-inch AA guns started by end of May 1943, with the guns being taken on loan from No. 1 AATC. After about a month, on 25 April 1943, the 9 B.O.s, 53 BOR, 5 VCOs, 587 IORs and 100 followers embarked at Karachi for Iraq, reaching Basra on 30 April, coming under the command of 40 Indian Infantry Brigade for administration and 4th AA Brigade for Operations and Training.[3]

The regiment stayed back in the Iraq when 4 AA Brigade, RA, returned to Middle East to beef up the run-down AA defences.

The regiment-less 73 HAA battery later moved to Assam in 1944 with the 73rd HAA Battery having moved to Ceylon in September 1944 and coming under command of 24th AA Brigade, RA.

CEYLON

As part of India Command, Ceylon (now Sri Lanka) was not a separate theatre nor a different campaign per se, but due to its geographical disposition, is being discussed separately.

In 1941, Ceylon had only one AA battery of the Royal Marines equipped with 3-inch guns. The anti-aircraft defences were reinforced by the raising of a Ceylonese AA regiment, and the arrival of three British AA regiments by March 1942. The AA units were placed under the Headquarters, AA Command, at Colombo, with guns deployed at Colombo, Trincomalee, Puttalam and Ratmalana. An early warning radar was available but the HAA regiment lacked sufficient GL sets. The AA defences were soon tested as the Japanese raided Colombo and Trincomalee on 6 April 1942. Over 60 Japanese planes, Navy Type 97 ('Kate') bombers escorted by 'Zero' fighters raided the targets in the two cities. Lack of warning made the task of the AA defences difficult but they did shoot down five aircraft over Colombo and six over Trincomalee.[4]

To strengthen the air defences, three more airfield were established, and an AA brigade formed on 9 September 1942—the 23rd AA Brigade in Colombo. Initially formed with Royal Artillery regiments and one AA regiment of Ceylon Garrison Artillery, it was responsible for the AA defence of the entire island. Its first Indian AA component was the 9 (Punjab) HAA Regiment which had been formed on 1 June 1942 at Lahore with three batteries (24th to 26th HAA Batteries). Originally the 14/8 Punjab, it was one of the seven infantry battalions that converted to AA regiments in 1943. One of the batteries (the 25th HAA Battery) was disbanded in April 1943; at the time the regiment had been ordered to move to Ceylon. The regiment reached Colombo in April 1943 and its two batteries were deployed at Colombo and Ratmalana.[5]

In February 1943, the 24th AA Brigade was formed at Trincomalee, and was made responsible for the AA defence of Northern Ceylon. Its first Indian AA Regiment was the 14th LAA Regiment that arrived in April 1943.[6]

By mid-1943, the Japanese air threat was limited to high-level reconnaissance sorties but the readjustments and upgradation of air defences continued. The 9 (Punjab) HAA Regiment was moved to Trincomalee in October 1943, coming under the command of 24th AA Brigade. To mislead the Japanese, a number of dummy Bofors anti-aircraft guns and radar towers were installed. These dummies were produced by Walker and Company, which otherwise produced machinery for tea and rubber industries during peace time.[7]

There were not many Japanese aerial intrusions except for occasional sorties towards Trincomalee. One of the tactics adopted by the Japanese was to try and deceive the radars by the use of 'chaff'. The war diary of 24 AA Brigade for 11 November 1943 notes how a 'Japanese plane dropped one piece of metal coated paper 60 × 1¼-inch, similar to the type used by our bombers over Germany to confuse enemy radars'. In another incident, the brigade notes, 'target pickup by RDF at 40,000 yds. *Multiple target effect* seen and "*beating*" repeatedly'. Such incidents of Japanese planes using chaff or resorting to Electronic Warfare (EW) tactics were repeated later as well, though there was never a serious attempt by the Japanese to overwhelm the defences or to carry out a raid under the cover of such EW measures.[8]

The next Indian AA regiment to arrive in Ceylon was the 13 Indian HAA Regiment that reached Colombo in March 1944 and came under command of 23rd AA Brigade. The regiment was deployed at Colombo. Later, the 102 Composite AA Battery on arrival ex Addu Atoll became part of the formation till its move to India in September 1944 for disbandment.[9] The 102 Composite AA Battery had handed over its 40-mm. Bofors guns on arrival but was later informed that the decision on its disbandment had been deferred and it would move to India instead. On reaching India in November 1944, it was reorganised as a LAA battery with battery HQ and three troops, received Bofors guns from 9 (Rajput) LAA Regiment and was deployed at Jamshedpur.

The 24th AA Brigade had two more Indian AA batteries; the 103/17 Indian LAA battery that joined the formation in August 1944 from India, followed by the 73/25 Indian HAA Battery from Iraq in September 1944. Headquarters 23rd AA Brigade closed down at Colombo in December 1944 and took over the responsibility of 24th AA Brigade at Trincomalee on 3 January 1945, the latter required to move to Assam. As per the orders issued by HQ ALFSEA, the 24th AA Brigade was to be formed as AGRA (AA), 14th Army, with one HAA and two LAA Regiments and one AAOR in mobile role. The brigade HQ did move to India—to Jorhat in January 1945 and took under command 6 Punjab LAA Regiment, awaiting its move to the Arakans, as its role was changed and the formation was now tasked with supporting the operations of the 15th Corps.[10]

In January 1945, with one of the brigade headquarters moving out, the air defences were readjusted, with the 3rd Searchlight/ Light AA Regiment CGA taking over the guns and gun positions of 76th LAA Regiment, RA. It was renamed as 3rd LAA Regiment, CGA. Some of the searchlights were taken over by the 103rd Indian LAA Battery of the 17th Indian LAA Regiment, becoming a composite searchlight/LAA battery though it was not renamed per se. This was first and only instance of an Indian AA unit operating searchlights.[11]

Another change in the operational role for the battery was the coastal defence task assigned to it with emphasis on 'anti-Motor Torpedo Boat' operations for which the battery did regular firing practice on a target towed by a Motor Torpedo Boat (MTB).[12]

The remaining period was rather uneventful with the Japanese never posing any serious air threat to the island.

MALDIVES

Early in the Second World War, the threat posed from the Japanese military push into Malaya and the Indian Ocean prompted the Royal Admiralty to plan for the relocation of the Eastern Fleet to a fallback safe harbour. In 1941, the Royal Navy began searching for a safe, deep anchorage, in a suitably strategic position in the Indian Ocean, where a naval base could be established. It was felt that Addu in the Maldives met these requirements and a party consisting of Fleet Air Arm, Royal Artillery, Royal Engineers and Royal Navy personnel were sent to the islands to reconnoitre them and ascertain its suitability for being developed as a naval base. Soon after, work began in August 1941 to secretly develop a fleet anchorage and base known as 'Port T'. It included an airfield which was developed at the Island of Gan.[13]

The anti-aircraft defences included a composite AA Battery that was provided by India. The 9 Heavy AA Battery (less two troops) and two troops of 11 Indian LAA Battery were sent to Addu Atoll in February 1943 as the first complement of AA artillery to be deployed there. Later, 102 Light AA Battery, that was raised in April 1943 along with 101 and 103 LAA Batteries as part of the 17 Madrassi LAA Battery, was despatched to Addu Atoll in November 1943 as an independent battery, serving as part of Island Fortress from 25 November 1943 to 6 May 1944.[14] The battery had 2 gun sections with 3.7-inch heavy AA guns and one LAA troop with four 40-mm. Bofors guns. The guns were deployed in a dual role—AA and coastal defence and came under the Commander Royal Artillery for operational control. The radar operators were sent to Ceylon for training on the GL set while the gunners were trained in situ. Thereafter, the 102 LAA Battery moved to Colombo, coming under command of the 24 AA Brigade, where it served till September 1944.

Despite it being ready to be used as an alternate fleet anchorage, in the eventuality of the Eastern Fleet withdrawing from its primary bases, the harbour was never used for its intended purpose and the Fleet withdrew to Kilindini instead when that eventuality did occur in May 1942. Addu Atoll's importance waned thereafter (the middle of 1943), as it became clear that it was very unlikely that Port T would be needed as a major fleet anchorage or a main servicing base. The RAF station also never fully commissioned the crew complement with only a small number of personnel being located at the base. It was later redesignated as a refuelling facility and RAF observation centre.

The AA defences, however, remained at Addu Atoll for the duration of the war and, sadly enough, the AA units located there do not merit even a footnote in the history of AA Artillery.

ZIPPER—THE OPERATION THAT WAS NEVER TO BE

As SEAC prepared for the last campaign to defeat the Japanese, one of the operation planned was code named Zipper, to capture either Port Swettenham or Port Dickson, Malaya as staging areas for the recapture of Singapore in Operation Mailfist. The operation was to be carried out by the 34th Corps which had the 8th Sikh LAA and 9th Rajputana LAA Regiments as corps troops. The two regiments were located at Bombay in early 1945, to train for the amphibious operation at the Combined Operations Training Centre. HQ 9th AA Brigade had been tasked to coordinate the training of the AA regiments, having shed its British AA regiments after conversion to medium regiments.[15]

The 9 AA Brigade, RA, had handed over responsibility of its area to 3 Indian AA Brigade on 16 April 1944 and moved to Ahmednagar and then to Pashan near Poona for training on medium artillery procedures. With this, the brigade ceased to be part of 33 Corps and came under HQ Southern Army as GHQ Reserve. During its stay, 53, 56 and elements of 63 plus 70 HAA regiments of Royal Artillery converted to 85, 86 and 87 Medium Regiments and 59 and 60 LAA Regiments, RA, were disbanded. The 9 AA Brigade moved to Secundrabad in September 1944 with 85, 86 and 87 Medium Regiments.

The role of the brigade changed in January 1945 and the first regiment to join was the 9 Rajputana LAA Regiment that arrived in January 1945. The regiment collected 18 40-mm. AA SP guns from COD Kirkee and 36 40-mm. AA Mk I guns from AATC Mehgaon for its new role. It was joined by the 136 and 137 AAOR (Mobile), 2 Indian HAA Regiment and 1 Indian Smoke company. The brigade was now to train for mobile and combined operations at the Combined Training Centre at Khadakvasla, outside Poona. In May 1945, Brigade moved to Gulunche and the training continued on amphibious operations.

Amphibious operations were something new for the Indian anti-aircraft regiments. From training to mount their guns on Morris trucks for embarkation and disembarkation to tactical siting of the guns, and control and reporting system on a beachhead, the regiments had to go through the entire regime. Captain Arthur Finn, one of the British officers posted to the 9th Rajputana LAA Regiment, recollects:[16]

> One of the first things I asked for at Deolali was a posting East of Dacca for the very good reason that rates of pay were higher if you were stationed there. This, however, produced no response from the Army and I landed up in Poona with a permanent posting to a unit that was booked for the inevitable invasion of Malaya. They were an anti-aircraft unit (9th Rajputana Light Anti-Aircraft Regiment) but with 40-mm. Bofors light guns, not the heavy guns I had been used to. Their task was to put these guns, self-propelled, on the back of Morris trucks down on the invasion beach and protect the invading troops from low-flying Japanese aircraft.
>
> So here I was with guns I knew nothing about, with troops whose language I couldn't speak and who had already completed their combined operations training. The prospect was not attractive. However, some progress was made learning Urdu and some knowledge of the guns added to a hope that things might turn out all right in the end.

To cater to the new role and tasks, a new scale of defence was laid down by the brigade vide its Policy Instructions No. 3 of 14 May 1945. Each brigade beach was to be provided one HAA and one LAA battery for anti-aircraft defence while the airfields were to be given two HAA and one LAA battery. The layout of the anti-aircraft defence was also altered to suit the requirement with a linear pattern recommended for deployment on the beach. Use of smoke, to be provided by 1 Indian Smoke Company, was recommended for both mobile and static defences.

The training and preparation for the invasion continued even as Burma was retaken, though the bombing of Hiroshima and Nagasaki leading to the surrender of Japan in August 1945 meant a change in plans. Operation Zipper was never fully executed and the 34th Corps was now to carry out the operation as a reoccupation force and not as an invasion force. One battery of the 9th Rajputana LAA Regiment was to accompany the reoccupation force for any eventuality, and the AA guns, trucks and equipment were waterproofed for the operations and sent to Bombay to be loaded on the ships.

As Captain Finn recalled:[17]

> We had waterproofed the trucks and equipment, had sent them to Bombay harbour to be loaded on the ships and were ourselves preparing to embark soon with the date for the invasion set, as I later found out, for 9 September. The Hiroshima and Nagasaki bombs ended the war on 15 August, about a week before we were due to set sail. In the end we sent one battery of about 250 men who were all Punjabi Mussulmen (Muslims) to do the landing and the two Rajput batteries stayed in India.

In early September 1945, the convoys of 34 Corps set sail for the designated landing beaches in the Port Swettenham-Port Dickson area and the first landing was carried out at Morib beaches on 9 September by the 25th Indian Division and at beaches of Sepang by troops of the 23rd Division. No Japanese resistance was encountered. The operation was a success—the last operation in which Indian AA artillery participated in 1945.

NOTES

1. N.W. Routledge, *History of the Royal Regiment of Artillery: Anti-Aircraft Artillery*, Brasseys (UK), London, 1998, pp. 198-200.
2. Ibid.
3. War Diary of 25 Indian HAA Regiment for the period March-April 1943.
4. John Clancy, *The Most Dangerous Moment of the War: Japan's Attack on the Indian Ocean, 1942*, Casemate Publishers, Philadelphia, 2015, pp. 83-8.
5. War Diary of 9 (Punjab) HAA Regiment, April 1943.
6. War Diary of 23 AA Brigade, RA, for April 1943.
7. Ashley Jackson, *The British Empire and the Second World War*, Bloomsbury, 2006, p. 3088. War diary of 24 AA Brigade for 11 November 1943.

8. War Diary of 24 AA Brigade, File No 601/2096/WD held with History Division, Ministry of Defence, New Delhi.
9. War Diary of 102 Composite AA Battery, September 1944.
10. War Diary of 24 AA Brigade, RA, for January 1945.
11. War Diary of 103 Composite AA Battery for January 1945.
12. War Diary of 103 Composite AA Battery for January-February 1945.
13. Jackson, *The British Empire and the Second World War*, pp. 331-2 and General Sir Martin Farndale, *The Far East Theatre, 1941-6*, Brassey's, London, 2002, pp. 8, 113.
14. War Diary of 102 Composite AA Battery for November 1943–May 1944.
15. War Diary of 9 AA Brigade, RA, for January 1945.
16. Judy Dean, *An Easy War in the Royal Artillery (Part 2)*, WW2 People's War at BBC, London, 26 May 2005 accessed at https://www.bbc.co.uk/history/ww2peopleswar/stories/05/a4119905.shtml on 10 July 2019.
17. Ibid.

CHAPTER 9

After the War

The 26 Infantry Division marched into Rangoon unopposed in an almost anti-climatic manner, but the fall of Rangoon was by no means an end to the war. Pockets of Japanese resistance remained, though it certainly was the end of air raids and bombing runs. The role of anti-aircraft regiments was almost over and it was no surprise that the end of war in Burma brought a pleasant reaction to all ranks, as also a sense of disappointment for a large number of anti-aircraft units for they never had an opportunity to participate in operations. This was more so with the regiments and batteries deployed in India in static role, as the war diary of the 4th Indian LAA Regiment notes:[1]

> Morale is high. Cessation of hostilities produced a pleased reaction in all ranks, only slightly subdued by the fact that the regiment, after all the hard training, had never had a chance to show itself off.

The regiments deployed in Burma and Assam were still busy with their operational duties as war with Japan was still not over. Plans were afoot for an invasion of Malaya and Singapore, and later of Japan. This, as the strategic bombing of Japan was still underway and the preparation for the dropping of the atomic bomb was being made in secret. Operation Zipper, the planned amphibious operation against the Japanese at Malaya, was of more immediate concern as two of the Indian AA regiments had been earmarked for the same. They were located at Bombay, training for the combined operations under HQ 9th AA Brigade, RA.

The Japanese surrender meant that the war was over for the British gunners with Indian regiments and they were now looking forward to be repatriated to England. The first to be so identified were the BOR with regiments in India—one of which was the 9th Rajputana LAA Regiment. The reaction of most gunners, a mix of joy and apprehension, and the need to train for future employment, is summed up in the record maintained by a British gunner in his personal diary:[2]

> 17th August 1945 brought the news that the war with the Japanese was over. The whole camp rejoiced as thousands of motor horns relayed the good news across the vast site. Within days, changes were taking place in everyday routine. Normal gunnery training continued in the camp, but all the British troops were invited to take part in vocational training in preparation for their impending return to civilian life. Courses for the British troops were being organised, both inside the camp at Gulanchi and in other camps in the greater Poona area.

The Indian regiments in Burma continued with their routine—and a new task of maintaining internal security and peace. The operation against the Burmese rebels were officially called 'anti-dacoit' operations with the 1st Indian LAA Regiment being employed on one such 'anti-dacoity' operations in Magwe area in September 1945, taking over the responsibility from the 4th Jammu & Kashmir Battalion. It deployed a party of 2 officers, 1 VCO and 45 men in support of police operations against a 'gang of dacoits led by San Din'. The anti-dacoity operation ended on 18 October 1945 with the killing of San Din, and the party of the 3rd Indian LAA Battery reverted back to the regiment.[3] But this was not the end of anti-dacoit operations as the 2nd Indian LAA Battery was soon used for supporting police action against another dacoit, Kywet Thoe, in Shwedaung.[4] These operations were a new experience for the anti-aircraft gunners but they conducted themselves with élan and quiet efficiency, just as they had conducted their previous operations. In November, the 1st Indian LAA Regiment handed over all local duties to the Chin Hills Battalion and moved to Rangoon, pending its move back to India; reaching Calcutta on 28 December 1945. It reached back four years after its first arrival in Rangoon, almost to the day.[5]

DEMOBILISATION

The first major step in fixing demobilisation targets was taken in January 1944 when the Commander-in-Chief instructed the Chiefs of Staff Committee to prepare a paper defining the size and composition of the defence forces that would be necessary for India to maintain after the end of the war. This was to be prepared on the assumption that no constitutional changes had occurred in India which would affect the status or composition of the armed forces. The report by the Chiefs of Staff was submitted in March 1944 but was used on as a foundation for further detailed examination by a Reorganisation Committee (India) set up in November 1944 that submitted its final report in October 1945. Meanwhile, an interim report was submitted by the committee in April 1945, giving a tentative proposal on the size of the post-war army.[6]

The Chiefs on Staff Committee on its part gave out its proposals for an immediate post-war period setting the upper limit of the army and the ultimate post-war army delineating the lower limit of the armed forces. Budgetary constraints also came into play as it was expected that for maintaining even the lowest strength of the armed forces, there would be requirement of a sum of Rs. 130 crore while the finance department was planning on a total of Rs. 70-75 crore only being available.[7] The revised proposals were submitted by June 1945 which brought down the cost for the lower limit of the armed forces to Rs. 80 crore—a more financially acceptable preposition though it was a serious cutback and was in no way a military recommendation.

While the plans for cutbacks and demobilisation was being worked out, the General Staff Branch prepared a new plan for demobilisation which laid down the estimated size of the Army for the next eighteen months in three phases of six months each, starting October 1945. Part I of the plan—called the 'Plan 287'—was approved by the Commander-in-Chief on 4 October 1945, and subject to amendments issued later on, became the basis on which the demobilisation of the Indian Army was carried out.[8]

The end of the war brought in a round of reorganization in the Indian Army as per the post war target set by the General Staff. The first reorganization plan for the artillery was drawn up by the Artillery Directorate at the Army Headquarters and it was followed up by a 'Conversion Conference' held in October 1945 at the School of Artillery to formulate the structure of post-war regiment of artillery. According to the plans, the anti-aircraft branch was to shed some of its regiments, to be converted as medium, field and anti-tank regiments. In addition, two regiments were to be placed in suspended animation.

The demobilisation process of the Indian anti-aircraft artillery had been initiated in 1943 itself, with the 7th Sikh LAA Regiment being disbanded on 30 September 1943.[9] During the war, one more regiment, viz., the 13th Indian LAA Regiment was demobilized, on 31 March 1944, leaving 32 AA regiments—an equal number of HAA and LAA regiments, and three brigade headquarters.

On 4 June 1945, three regiments, i.e. the 8th Rajput, 9th Punjab HAA Regiment and 17th Indian LAA Regiment were placed in suspended animation. This was followed by conversion of one LAA and eleven HAA regiments to field/ medium/ anti-tank units as part of the reorganisation plan for artillery to rationalize the number of artillery regiments based on the recommendations of the Conversion Committee of the Artillery Directorate. While one light AA regiment was converted as a field regiment, three HAA regiments were converted to medium, six to field and two HAA regiments to anti-tank regiments in August 1945.[10] These regiments were:

Anti-Aircraft Regiment	*Converted to*
3 HAA Regiment	2 Medium Regiment
5 HAA Regiment	4 Medium Regiment
7 HAA Regiment	17 Field Regiment
9 HAA Regiment	12 Field Regiment
10 HAA Regiment	14 Field Regiment
11 HAA Regiment	15 Field Regiment
12 HAA Regiment	3 Anti-Tank Regiment

13 HAA Regiment	13 Field Regiment
14 HAA Regiment	3 Medium Regiment
20 HAA Regiment	9 Anti-Tank regiment
21 HAA Regiment	16 Field Regiment
18 LAA Regiment	11 Field Regiment

All the AAORs, that were raised during the war were also demobilized by August 1945 leaving a void in the anti-aircraft defences as there was no operations centre to coordinate and control the operations of the anti-aircraft regiments and batteries. The immediate need was thus for an Anti-Aircraft Operations Room each for the western and eastern regions. Accordingly, two new AA operations rooms, viz., Nos. 1 and 2 AAOR were raised on 22 February 1946.[11] With the planned demobilisation of the three AA brigade headquarters scheduled to be completed by August 1946, Headquarters 11th and 12th Army Group RIA (AA) were raised at Gulunche and Uruli, respectively, on 28 February 1946. Soon thereafter, Headquarters 2nd and 3rd AA Brigades were demobilized on 15 April 1946 followed by the HQ 1st AA Brigade on 31 August 1946.[12]

REORGANISATION

As the plans for demobilisation were being carried out, the Indian anti-aircraft regiments were reorganised to make the select regiments one class units. The 2nd Indian LAA Regiment shed its 7th LAA Battery composed of Punjabi Mussalmans, replacing it with the 8th HAA Battery of Ahirs from the 2nd Indian HAA Regiment. The battery was converted to LAA battery, and the existing 8th LAA Battery renumbered as the 9th LAA battery. This made the 2nd Indian LAA Regiment an all-Ahir regiment. The 7th LAA Battery went to the 1st Indian LAA Regiment, making it a Punjabi Mussalman regiment as it shed its 2nd LAA Battery of Punjabi Brahmins.

The 3rd Indian LAA Regiment was located at Pashan near Poona until end-November 1946 when it was moved to Rawalpindi to join the 10th Infantry Division. Before its move to Rawalpindi, the regiment shed a battery (the 13th LAA Battery) and despatched 183 OR to Mathura for release. At its new location, the regiment had a total strength of 12 Officers and 493 OR and was involved in routine training. It also took part in the 3rd Series Experimental Frontier Brigade Exercise from 8-12 December 1946.

In January 1947, the Indian artillery regiments were renumbered serially, to avoid duplication of numbers, with the Special Army order 10/S/47 giving out the new designation of the artillery regiments. Amongst the anti-aircraft regiments, the 2nd Indian HAA Regiment was renumbered as the 18th HAA Regiment, RIA, and the 6th as 19 HAA Regiment, RIA. The seven LAA regiments (the 1st, 2nd, 3rd , 4th, 6th Punjab, 8th Sikh and 9th Rajputana LAA Regiments) were renumbered as the 25th , 26th , 27th , 29th , 28th , 30th and 31st LAA Regiment, RIA.[13]

A further 187 OR of 27th LAA Regiment, RIA, were released in January-March 1947 and the regiment moved to Wah on 7 March 1947. It was deployed for internal security (IS) duties in Rawalpindi area during the month. With the release of more OR, the regiment placed 12 40-mm. AA guns (SP) and an equal number of gun-towing vehicles in mobile reserve. The regiment continued with IS duties during the period April to June 1947 and changed to Peace Establishment with effect from 4 June 1947.[14] After IS duties at Cambellpore, the regiment moved to Poona in September 1947 where it received reinforcements from the 29th LAA Regiment, RIA, and reorganized on a full three battery basis.

The 6th Punjab LAA Regiment was located at Hyderabad (Sindh) in early 1947 while undergoing conversion as a Para LAA regiment. The new designation, with effect from 1 April 1947 meant redesignation as the '28 (Punjab) Para LAA Regiment, RIA.' The regiment was planned to be equipped with 40-mm. bantam type AA guns by June 1947 and the men of the regiment underwent training at Chaklala, home to both the Army/Air Transport School and Paratroopers' Training School. The regiment also participated in an air dropping exercise at Chaklala but the future was uncertain as there were no suitable guns available. Just prior to the Partition of India, orders were issued for the regiment to move to Mathura where it was disbanded in

September 1947. The Quarterly Historical report of the regiment notes:[15]

> Morale of the men and the VOs is very low owing to regiment being disbanded and inspite of assurances given by the officers that every effort will be made to absorb every one of them into other regiments, they are not confident about the future.

The Punjabi Mussalman battery (the 19th LAA Battery) stayed back in Pakistan and the remaining Sikhs and Rajputs came to India. The officers and men were inducted into field and AA units though there was general dissatisfaction with the 'terms of service not attractive enough for good, capable men' as noted in its war diary.[16]

> Reaction of almost all Indian officers to new pay code is very unfavourable as they feel that the new pay code is not good enough to attract any capable man into the Army.

PARTITION

With the Partition of the country set to be formalized, the task at hand was the distribution of various units and formation between the two countries and the details of units to be allocated were to be decided by select committees. Based on the recommendations of the 'Army Sub-Committee for the distribution of artillery and engineer units' between India and Pakistan, India was to get two LAA regiments while Pakistan was allotted two HAA and two LAA regiments.[17] The imbalance between the anti-aircraft regiments to be divided between India and Pakistan was explained in the Partition Council meeting as reflective of the actual number of Muslim and non-Muslim units as they then existed in the Army.

To balance out the uneven distribution of anti-aircraft regiments, Pakistan was required to convert one heavy anti-aircraft regiment to a field regiment, and one light anti-aircraft regiment to an anti-tank regiment, and India was required to convert one field regiment to a heavy anti-aircraft regiment. This would result in India having three AA regiments (one HAA and two LAA regiments) and Pakistan two regiments (one HAA and LAA regiment each).[18] Among the training institutions of the Army that would be lost to Pakistan was the Anti-Aircraft Artillery School, Karachi.

The transfer of anti-aircraft regiments did not proceed as per the plans as further demobilisation took place, with the 19th HAA Regiment, the 28th (Punjab) Para LAA, and the 29th , 30th and 31st LAA Regiments being disbanded before the division of units took place.[19] This left only four anti-aircraft regiments—the 18th HAA Regiment and the 25th , 26th and 27th LAA Regiments.

In the end, the regiments transferred to Pakistan were the 18th Heavy Anti-Aircraft Regiment (with the 1st, 7th and 19th HAA Batteries) and the 25th Light Anti-Aircraft Regiment (with the 3rd , 6th and 7th LAA Batteries), with the 26th and 27th Light Anti-Aircraft Regiments being retained by India.[20] India did not convert a field regiment to HAA regiment; instead a HAA regiment was raised later, in 1948. The anti-aircraft regiments with India retained their numbers (they still do, to date) but the anti-aircraft regiments that went to Pakistan were re-numbered. The 18th HAA Regiment was renumbered as 5th HAA Regiment, RPA, and the 25th LAA Regiment as 6th LAA Regiment, RPA. The Pakistan Army retains these regiments to date, with the same battery designations.[21]

NOTES

1. War Diary of 4th Indian LAA Regiment for January 1945.
2. 'World War 2 Diary of Major Sidney Ernest Wright', Lennon Wyle, https://www.lennonwylie.co.uk/WW2DiaryWright2.htm.
3. War Diary of 3rd Indian LAA Regiment for October 1945.
4. War Diary of 1st Indian LAA Regiment for October-December 1945.
5. Ibid.
6. Bisheshwar Prasad, *Expansion of the Armed Forces and Defence Organisation 1939-45*, History Division, Ministry of Defence, Government of India, Pentagon Press, New Delhi, 2012, pp. 197-9.
7. Ibid., pp. 204-8.
8. Ibid., pp. 208-11.

9. *Statistical Review of Personnel*, Indian Army, vol. IV, 1 October 1945 to 31 December 1948, Compiled by Army Statistical Organisation (ASO/2), Government of India Press, New Delhi, 1948.
10. Maj. Gen. D.K. Palit, *History of Regiment of Artillery*, Indian Army, Leo Cooper, London, 1972, pp. 292-3.
11. *Statistical Review of Personnel*, Indian Army, vol. IV, 1 October 1945 to 31 December 1948.
12. Ibid.
13. Palit, op. cit., pp. 292-3.
14. Historical Report for Quarter Ending March 1947 of 3rd Indian LAA Regiment.
15. Historical Report for Quarter Ending September 1947 of 28th (Punjab) Para LAA Regiment.
16. Ibid.
17. Fourth Meeting of Partition Council, Allocation of Armoured Corps and Infantry Units, *Partition Proceedings*, vol. 5, p. 32. Sharmila Singh, 'Partition of the Indian Armed Forces between India and Pakistan', School of International Studies, Jawaharlal Nehru University, New Delhi, 1994, accessed on 14 March 2020 at http://hdl.handle.net/10603/14220.
18. Ibid.
19. *Statistical Review of Personnel*, Indian Army, vol. IV, 1 October 1945 to 31 December 1948.
20. Ibid., p. 428.
21. Maj Gen Shaukat Raza, *Izzat-O-Iqbal*, School of Artillery, Nowshera, 1980, pp. 16-19.

Conclusion

In 1943-4, India had the second largest concentration of AA artillery in the world, second only to Great Britain. It was one of the most extensive expansions ever for the Indian Army, from just one regiment in 1940 to thirty-three AA regiments in three years.[1] The beginnings of the Indian AA Artillery were modest, with the first AA regiment being raised in October 1940 at Karachi, using the manpower of the 8th HAA Battery, RA, and drawing in the experienced VCOs and NCOs from the Indian Mountain Artillery. With no base of its own, the initial raisings used the pre-war class, mainly the Punjabi Muslims and Sikhs, but as this recruitment base was not large enough to meet the demands of all arms, the AA artillery looked at the South Indian classes for recruitment and soon enough, the majority of units were made up of 'Thambis', as the Madrassis came to be called.

In 1943, no fewer than five light AA and eight heavy AA regiments (Nos. 10, 11, 12, 17 and 18, and Nos. 11, 12, 13, 15, 16, 20, 21 and 25, respectively) were formed of Madrassis only and, by the end of the war, Madrassis preponderated over all other groups in the Indian AA artillery.[2] There were apprehensions in the beginning about the suitability of recruiting the South Indians but the performance of the 'Thambis' put the unfounded fears to rest.

The first regiment to go into combat was the 1st HAA regiment which sailed for Singapore in July 1941 while still under training. It was tasked to take on a mix of AA guns at Singapore, from 3-inch HAA to 40-mm. LAA and static naval AA guns, most of which the gunners had not used or been trained on. To add to the difficulties, the regiment was augmented by a draft of British gunners, leading to language and communication problems.

Another challenge was a frequent change in tasks and deployment areas for the batteries and troops. Inadequate air cooperation and shortage of training ammunition meant that the majority of the gunners were not yet trained fully when the battle was joined in December 1941.

The Japanese Air Forces were far better and much more effective than expected during the Malaya campaign.[3] The large Vics of 18 to 27 planes were not only accurate but devastating in their attacks on targets on the Singapore island. The tactics of continuing with the attack in face of AA fire meant that normal deterrent fire not enough to ward off a raid. All this, and the constant shelling by the Japanese with ground assaults, made the task of AA gunners all the more difficult.

And yet, one DSO, one MC and two IDSMs, with the scores killed and injured in the battle of Singapore area, bear testimony of the valour and sacrifice of the regiment, aptly, the *'First'* in all respects. With the regiment 'lost' in Singapore, it was never revived again. What is generally forgotten is the contribution made by the 5th Indian LAA Battery that had gone to Singapore with the 1st Indian HAA Regiment. This was the first Madrassi AA battery ever raised and it was a torchbearer in so many ways. The names of its fallen soldiers on the memorial at Singapore are the last of the legacies of the battery that led the way for the other Madrassi AA regiments and batteries to come.

RAISING OF UNITS

Raising of units was never a matter of only getting the required number of recruits together to form batteries but also of training them. The challenge for the training centres was to mould 'extraordinarily raw recruits', taken from a largely rural society bereft of access to modern technology, and turn them into good, technically proficient soldiers and they did this admirably for never were the Indian anti-aircraft gunners found wanting. The 1st HAA Battery was deployed even when it was being formed, as was the 1st HAA Regiment that went overseas to Singapore. Both acquitted themselves well during the war.

During the initial training, the common practice was to form groups of recruits under British NCOs, who acted as their instructors as well as their mentors. While the officers were required to learn Urdu, the British NCOs were not compelled to learn any Indian language, although they were expected to encourage the Indian troops to practise their English as much as possible. The outcome, of course, was that the NCOs found themselves learning smatterings of several languages from the Indians. The requirement for Indian gunners was only to learn primary English which was taught during the basic military training, as all orders were given in English.

Firing was the all-important practice for all gunners, in a way the culmination of training and a test of the standard of training of any unit. Related to this was the laying practice that the gunners carried out, preferably on live targets. There was only one Anti-Aircraft Cooperation Unit, No. 22 AACU, that had a flight each at Karachi, Bombay, Calcutta and Katni.[4] Even with 60-odd aircraft, the AACU was always hard-pressed to ensure the required air cooperation for the anti-aircraft regiments, and at times aircraft could not be provided to the units. The problem was aggravated by the deployment of regiments in batteries and troops, spread all over the country. To overcome this, and to have centralised training, the Eastern and Southern Armies had their Mobile AA Schools.[5] The batteries were sent to the training schools in rotation, with at least one battery from each brigade always undergoing training.

The deployment of the batteries and troops at the airfields did help in the laying practice, but as the flight profile of the deployed aircraft was not the same as that of any raiding aircraft, it was not a perfect solution. Still, it did help. For firing, the most common method was the use of drogues. They were towed by aircraft as varied as Waptis and Hawker Henley.

To make up for the non-availability of drogues, the regiments and batteries tried out all possible measures—from firing on the shadow firing to firing on live bursts. 'Below Barrage' and '18 degree offset' were two common methods of firing, when the guns fired at a point offset from the location of the aircraft.

Shadow firing was more commonly carried out by batteries deployed on sea coast when the guns would fire at the shadow falling on the sea waves.

The firing standards of Indian anti-aircraft regiments were generally good. Two examples would suffice. The first is mentioned in the personal diary of Major Sidney Ernest Wright, an officer with the 15th Indian HAA Regiment:[6]

> From now onwards, the days were occupied in training—technical, and all the other branches. By January, 42 and 43 Batteries were ready for Practice Camp. Each unit spent a week at the AA School in Karachi, firing each day. Very good results were obtained, and right sleeves were shot down. Satisfactory reports were issued by the School to GHQ on the unit's progress. The shooting was certainly of a higher standard than one would expect from a young Indian regiment. It was more consistent and accurate than the 8th Belfast when I saw them fire at Practice Camp. It now remains to see what will be their reaction to action against enemy targets. After this camp, the Batteries return to Malir and concentrate on the weaknesses which were found in the firing. After a further month of training they return to AA School, and take the final Practice Camp before mobilization of the regiment.

What is of special interest is that the newly trained Indian AA batteries had achieved better results than a battle hardened Royal Artillery regiment. As Wright mention, the shooting was 'more consistent and accurate than the 8th Belfast'. The second incident was during the retreat from Burma in 1942. The 3rd Indian LAA Battery was one of the Indian anti-aircraft batteries providing air defence to the retreating troops. While the 'B' troop of the battery was at Taungdwingyi, a Japanese air raid came in. The gunners opened up, shooting down the aircraft with the first four rounds.[7]

Training was not confined to firing practice only. The regiments constantly strived to train their men, to mould them into good gunners. In one instance, the regiment moving to a new location came to know that the guns being handed over to them were a different Mark (Model) than the ones they had previously manned, the main difference being in the auto-loader of the gun. The battery commander, not wasting any time, organised classes in the train as the battery moved to its new location.[8]

The training of anti-aircraft gunners was a continuous process, repetitive to a point but necessary to hone the skills required to shoot down a fast moving aircraft with split-second reaction time. The requirement of manning an AA gun and radar was far more complex than handling a rifle but the Indian *sipahi* did well. The training had a lot to do with this. The initial bonding with officers and NCOs as well as using advanced training methods helped.

One of the challenges was to teach aircraft recognition to the gunners who had previously never seen any planes in their life. Charts alone were of not much use and it was aircraft models that proved to be of greater use in teaching the jawans about the techniques of recognising different planes.[9] But modern training aids were not available at all locations and innovative methods had to be adopted by the battery and regimental commanders. Dome trainers for L/60 Bofors gunners were used only at select training centres and schools.[10] Otherwise, use of 'shadow firing' and firing on meteorological balloons using LMGs were some of the techniques adopted by the units.

OFFICERS

Officers that came to Indian Anti-Aircraft Regiments were a mixed lot. The initial lot were mostly those who had served with Indian troops and had knowledge of Urdu, but as the anti-aircraft artillery expanded, such officers were hard to find, and an increasing number of officers were just being produced by Cadets' Wings and Officer Training Schools. The 1st HAA Regiment received a draft of thirty officers while at Singapore, twenty-four of whom were subalterns with no knowledge of Urdu (or any Indian language) and no experience of Indian troops. This did create some adjustment problems, but they soon learnt to tide over them.

Transfer of officers from infantry and other arms was common during the early years. When the seven infantry battalions converted to anti-aircraft regiments in 1942, the battalion officers were transferred to anti-aircraft artillery, making it one of the largest such transfers of officers. Lt Col V.T. Smith was one such infantry officer who became an anti-aircraft gunner rather late in the day. He commanded the 7th (Rajput) HAA Regiment and went on to command an AA brigade during the war and was the senior-most AA gunner at the end of the war.

Knowledge of Urdu, or any other Indian language, was a must. Special classes were conducted for the same, for all newly posted officers. The VCOs also played a role in this, for they were the all important link between the officers and the men. Some of them had good knowledge and understanding of English, with a handful even qualified as trained interpreters but they ensured that the officers learnt the local language. As one Jemadar told 2/Lt Lancelot Nash of 3rd LAA battery, 'if I had told you that I knew English, you would have never learnt Punjabi.'[11]

The link language in the units was not always Urdu as the class composition of the regiments varied. The first few regiments were made up of recruits from Punjab—Muslims, Sikhs and Brahmins—and the preferred language in quite a few batteries was Punjabi rather than Urdu though Urdu remained the official language. Similarly, in the Madrassi regiments, use of English was common and most of the officers did pick up a working knowledge of one of the south Indian languages.

The communication problem was greatly eased when the Indian officers started joining the anti-aircraft units. The arrival of Indian officers was always a matter of joy, and relief, for the commanding officers, as the regiment then had 'someone who can talk to the men in their own language!'. Amongst the first batch of these officers were 2nd Lieutenants Saheb Singh Kalha, Amreek Singh and Hartaj Singh who were commissioned into the Indian Anti-Aircraft Artillery on 22 December 1940. Major Gurbux Singh was the brigade major of the 3rd AA brigade during the war, the first Indian officer to serve as a staff officer in an anti-aircraft formation, though the first Indian officer to command an anti-aircraft regiment had to wait till 1947 when Lt Col J. Raj Bahadur assumed the command of the 27 LAA Regiment.

Barring an odd case, the Indian officers did well. One such example of the 'odd' officer is of 2nd Lieutenant Mehar Dass of 3rd Indian LAA Battery who served in Burma in early 1942. He was last seen jumping off a launch to go back to his troop and try to retrieve or destroy a Bofors gun that had been left behind. The war diary of the battery mentions the incident in glowing terms but does not give any details of his whereabouts after this incident. While a gunner of the battery did mention that he saw Lt Dass in Rangoon at a later day working in a labour camp. As Lt L.L. Nash of his battery later recounted his experience during the Burma campaign, 2nd Lt Mehar Dass subsequently joined the Indian National Army (INA) and even took part in an operation against the 1st Indian LAA Regiment.[12] This was the only instance of an anti-aircraft officer joining the INA.

MADRASSI UNITS

The 5th Indian LAA Battery was the first Madrassi anti-aircraft unit to be formed. It proceeded to Malaya in November 1941 and was one of the units that was lost after the fall of Malaya. The next Madrassi AA batteries to be raised were the 6th Indian Heavy AA in June 1941 and 9th Indian Light AA in October 1941. The year 1942 saw the formation of the 3rd, 4th and 5th LAA and the 6th and 10th HAA Regiments, which were composed entirely of Madrassis, while the 9th Punjab HAA Regiment included one Madras battery, and one-and-half batteries of Madrassis were in the 5th Indian HAA Regiment.

Finally, in 1943, no less than 5 Light AA and 8 Heavy AA Regiments (Nos. 10, 11, 12, 17 and 18, and Nos. 11, 12, 13, 14, 16, 20, 21 and 25, respectively) were formed of Madrassis only, so that by the end of the war, Madrassis preponderated all other classes in the Indian artillery. Of the two LAA regiments that were retained after Partition, the 27 LAA Regiment was a South Indian Classes unit, and remains so till date.

The expansion brought in some challenges that were not initially planned for. The tapping of recruitment base in south India was a well-thought-out strategy to overcome the limited base of Punjab but what was not considered till late was the shortage of drivers and mechanics that would arise as the units started getting formed.

There were no mechanised units in Indian artillery at the outbreak of the war, and the training of drivers and driver-mechanics was never given the due importance when the training centres started expanding. Seemingly a minor issue, but became of critical importance as the anti-aircraft units started getting mechanised from 1943 onwards, with their 'semi-mobile role'. The time taken to train a driver-mechanic was almost double the time taken to train a gunner, and this started affecting the availability of the former with the units. The first establishment to introduce training of driver-mechanics for Indian artillery units was the FATC, Mathura, in 1941-2 followed by the No. 2 AATC. Similarly, availability of signallers was always at a premium and many a unit carried out training of suitable men as signallers and giving them additional allowance, using regimental funds.

JAPANESE AIR THREAT

There was no separate air force in Japan with both Army and Navy having their air arms, independent of each other.

The Imperial Japanese Army in its modern form dates to 1871 and the army aviation started in 1911. In 1940, the Japanese Army Air Force had five air divisions, three of which were based in Japan with two in Manchuria. The Japanese Army Air Force employed its 3rd Air Division during the Malaya campaign with about 250 aircraft while the 5th Air Division was used during the Burma campaign. The 22nd Air Flotilla of Imperial Japanese Navy with 110 aircraft operated from three bases during the campaign in Malaya.

Massed attacks were the cornerstone of the tactics adopted by the Japanese Air Forces during both campaigns. The plane formations used similar methods—of an advance reconnaissance aircraft (or a small formation) to provoke the anti-aircraft defences to open up, giving away not only their location but also the extent of the VA. This was followed by the larger formation which carried out the main attack(s). The number of planes in a formation varied, and were generally in multiples of 9. Formations at different altitudes, the advance formation at lower altitude with the main formation coming in higher to stay out of anti-aircraft gun range, was another method adopted by the Japanese planes. Direct attack on to the anti-aircraft guns, after having made them open up, was occasionally tried out but it was not a regular tactic of the attacking formations.

The use of radar avoidance techniques adopted by the Japanese were not always successful, partly because of the nature of these tactics and techniques itself. Simultaneous attack from multiple directions to overwhelm and confuse the radar was an effective technique, as was taking cover of hills and valley, but 'coming low' did not work for the Japanese. In a mistaken belief that the radar could not pick up planes at low level (*not* nap of the earth or terrain hugging), at times the Japanese carried out raids at low level and were effectively detected by the radars, though this was not a common practice and was most likely carried out in a handful of cases only.[13]

Repetitive use of these tactics made them predictable and the anti-aircraft defences stopped engaging the reconnaissance planes, unless they were well within their range and a hit was assured.

Use of electronic warfare measures was noticed during some of the raids, chaff being the common form. Flying at low altitude and taking advantage of terrain, like flying in valleys to evade detection by radars was part of the training of Japanese pilots. Another aspect that was drilled in them was to maintain the formations even in face of hostile anti-aircraft fire. During the earlier campaigns, it was observed that the attacking formations seldom broke up, however, intense the AA fire, but in the later campaigns, the aircraft formations were seen breaking up as the anti-aircraft defences opened up. One reason for this change could be the standard of pilots, as newer and inexperienced pilots joined the frontline squadrons.

One of the unexplained failures of the Japanese Air Forces was not exploiting their initial advantage and carrying out more raids and attacks against the airfields and other installations in Assam and Bengal. With over 80 airfields spread over north-east India, a major air campaign could have crippled the Allied war effort, and it was in the realm of possibility in 1942-3 as the anti-aircraft defences were still in the process of building up.

ADMINISTRATION

The challenges of administration of the Indian anti-aircraft regiments were the same as that of other Indian units except for an added problem, that of deployment in batteries separated from the regimental headquarters and, at times, at great distance from the nearest military cantonment or formation. Distance may make the heart grow fonder but it was not so with the anti-aircraft batteries. The headquarters often issued directions that were not practically implementable by the batteries, like digging slit trenches around the gun pits when the entire area was often waterlogged.

The weather and administrative failures often led to situations where the jawans had to face unwarranted hardships. In one of the instances, which would have occurred in many a regiment, the battery commander of the 3rd LAA Regiment at Matmansingh notes:[14] 'No blankets for the troops. The nights are cold and many men are suffering from chills, leading to fever...'

In other locations it was dengue. Another war dairy notes, 'sickness has been high during the last month—in some cases it is as high as 18 per cent on gun sites. The chief complaints are dengue and skin trouble.'[15] It led to a state where the manning of guns was affected and batteries had to borrow gunners from other regiments to continue the activity.

The routine affairs of a battery administration were, however, rather mundane. The war diaries mention the occasional court of inquiry to 'investigate into the circumstance in which Gunner XYZ absented himself without leave' or the case where the 'dhobi's tent mysteriously caught fire in the night of XXXX' but these were the exceptions to the routine. The Indian anti-aircraft gunners were a hardy and sturdy lot, better than the British Johnny in most aspects. 2/Lt Lancelot Nash, while describing the march back to India, mentions that the Indian gunners did well in comparison to the average British gunner. Another case was of

Gunner Mohd Amir who had been captured by the Japanese in Burma and put in a labour camp, making roads. He managed to escape and walk back to India, via., China. What was remarkable about this feat was that Khan did not know any other language except Punjabi. When asked, he simply said, 'I just walked steadily in the right direction.'[16]

Even otherwise, the Indian anti-aircraft gunners took pride in their profession, and their uniform. More than one regimental commander has mentioned in the war diaries and monthly reports that 'Indian gunners are better turned out than their British counterparts and are more disciplined' though there was obvious discrimination between the administrative facilities extended to British and Indian OR in the same battery.[17] An entry in the 3rd Indian LAA Regiment war diary for 25 January 1943 notes, 'Inspector of Catering visits and looks around British OR catering arrangements.' There is no mention of any inspection of an Indian OR's mess. Maybe it was never felt important enough or the 'caste' restrictions were such that any visit by an outsider in an Indian jawan's mess was not acceptable.

The difference in treatment meted out was in almost all aspects. The standing orders of the 3rd Indian Brigade specify that 'when an officer passes a sentry standing in the open, the sentry will salute the officer in hours of daylight. British troops will give a butt salute and Indian troops will give a shoulder salute.' Such orders were common during the period in Indian Army and though they may not have affected the operational efficiency of the 'mixed units,' they do point to the discrimination between an Indian jawan and a British Johnny in the same unit.

THE CAMPAIGNS

The self-pride and discipline of Indian anti-aircraft gunners was reflected in the performance of their units. From Malaya and Singapore to Iraq, the units served with honour. Most were deployed in their primary role of anti-aircraft defence, but a large number, specially the LAA regiments, were used in ground role by the field formations. Even the heavy AA regiments at times used their 3.7-inch guns against ground targets to effect, though this did create an unexpected result at times. The HAA regiment with the 23rd Indian Infantry Division wanted to try out the use of their 3.7-inch guns in an indirect role, more like a mortar. When they fired at the practice range, the round went straight up, and 'the gunners peered through their binoculars at the target end. There was a pause, a very large pause, before there was a loud explosion some way behind the gun.' Apparently, the round became unstable over 40,000 feet and its trajectory could not be predicted. That, thankfully, was the last of such experiments and no more trials were carried out.[18] The light AA guns did well in both anti-aircraft and ground role and were used with telling effect by the formations on the final push to Rangoon in 1944-5.

During the operations, from Singapore to Burma, the Indian anti-aircraft regiments won a DSO, three Military Cross, one Indian Order of Merit, one Order of British India, one MBE, eight Indian Distinguished Service Medals and twenty Mention in Depstaches. The first gallantry awards were won by Havildar Sham Lal and Gunner Balbir Singh at Tengah aerodrome, Singapore, in February 1942 while the last was won by Havildar Surat Singh of 2nd Light AA/A Tk Regiment.

While a lot can be said of the pioneers, it would suffice to say that they were honourable men, and they were gallant anti-aircraft gunners. They were the first Indian anti-aircraft gunners, and in the words of Major Charles MacFetridge, battery commander of the 3rd Indian LAA Battery:[19] 'It was remarkable how much the Indian anti-aircraft gunners achieved with so little.'

NOTES

1. A total of thirty-seven anti-aircraft regiment were raised. One was lost in Singapore in 1942 and one was demobilized early in September 1943. Two HAA regiments (Nos. 15 and 16) raised in October 1943 converted to LAA regiments with the same number thus bringing the maximum number of regiments to 33.
2. Lt Col E.G. Phythian-Adams, OBE, *The Madras Soldier, 1746-1946*, Government Press, Madras, 1948, pp. 193-4 and Memorandum

on 'Expansion of Indian Artillery 1939-1941', Prepared by Royal Artillery Directorate, GHQ, Delhi, History Division, Ministry of Defence, Government of India, New Delhi.

3. Lt Gen Percival, *War in Malaya*, Eyre & Spottiswoode, London, 1949, pp. 109-10.
4. Sqn Ldr Rana T.S. Chhina, *The Eagle Strikes: The Royal Indian Air Force, 1932-1950*, Centre for Armed Forces Historical Research, Ambi Knowledge Resources, New Delhi, 2006, pp. 29, 82.
5. 1st Indian AA Brigade War Diary for February 1942.
6. 'World War 2 Diary of Major Sidney Ernest Wright', Lennon Wyle, https://www.lennonwylie.co.uk/WW2DiaryWright2.htm, accessed on 21 July 2019.
7. As recounted by Lt Lancelot Lester Nash, 3rd Indian LAA Battery, at Oral History Collection of Imperial War Museum, London accessed at https://www.iwm.org.uk/collections/_item/ object/80020144, on 25 July 2019.
8. War Diary of 2nd Indian LAA Regiment for September 1943.
9. Tour Notes of Brigadier AD, Eastern Army, to 2nd Indian AA Brigade, July 1943.
10. The dome trainers for 40-mm. L/60 guns were not produced in large numbers to be given to each formation. Only 40 odd such trainers were produced, of which one was at AA School, Karachi, and the 24th AA Brigade, RA, had a dome trainer at Colombo AA Training School, but the details about other brigades are not available.
11. As recounted by Lt Lancelot Lester Nash, 3rd Indian LAA Battery, at Oral History Collection of Imperial War Museum, London.
12. Ibid.
13. 13th AA Brigade, RA Intelligence Survey, September 1943.
14. War Diary of 3rd Indian LAA Regiment for January 1944.
15. War Diary of 4th Indian LAA Regiment for November 1943.
16. As recounted by Lt Lancelot Lester Nash, 3rd Indian LAA Battery, at Oral History Collection of Imperial War Museum, London.
17. War Diary of 4th Indian LAA Regiment for January 1945.
18. Doulton, AJF, *Fighting Cock: Being the Divisional History of 23rd Indian Division*, Naval & Military Press, Uckfield, 2012, pp. 164-5.
19. Oral History as recounted by Major C.H.T. MacField of 3rd Indian LAA Battery. Accessed at Imperial War Museum, London at https://www.iwm.org.uk/collections/item/object/80020185, accessed on 25 July 2019.

Epilogue

Pakistan attacked India on 1 September 1965, launching Operation Grand Slam, with the objective to capture the vital town of Akhnoor in Jammu. The Indian Air Force was called in and it launched repeated sorties to stem the Pakistani advance but in the process lost four Vampire aircraft. The day did not end well for India. The next day, 2 September, the two air forces held back as the armies slugged it out, but there was an almost desperate need to counter the Pakistani Air Force and rein it in. With no Pakistani air raids over Army formations and installations, the Indian anti-aircraft artillery had not yet joined the battle.

The fighting on 3 September turned out to be different. The Indian Air Force shot down a Pakistani F-86 Sabre. Its first kill by the diminutive Gnat, piloted by Squadron Leader Trevor J. Keelor of No. 23 Squadron, IAF. The same day, a Pakistani mission over Tawi bridge was engaged by a 40-mm. L/60 section, with Havildar C. Perumal commanding one of the detachments. As the Sabres came in range, the Bofors opened up, hitting one of the Sabres and sending it spiralling down.

The Indian anti-aircraft artillery had its first kill, by Perumal of 27 Air Defence Regiment.

Six years later, on 4 December 1971, F-104 Starfighters of Pakistan Air Force attacked an Indian radar station near Amritsar on the western border. The war had broken out the previous day, with the Pakistan Air Force trying, and failing, to carry out a pre-emptive strike against Indian air bases and installations. This strike against the radar was a continuation of the previous day's operations.

Defending the radar was a battery of radar controlled 40-mm.

L/70 guns. Lance Naik Shreepati Singh was one of the radar operators.

As the Starfighters came in for the strike, one of them picked up the radar and directly attacked it. Shreepati Singh was outside, on the visual tracker, exposed. He locked on to the Starfighter with the visual tracker and continued to track it steadily even as he was hit and wounded. The guns, following the radar, found the target. Shreepati Singh kept tracking the target till it was shot down.

This was the first-ever Starfighter shot down by the Indian anti-aircraft artillery, by Shreepati Singh of 26 Air Defence Regiment.

These regiments were two of the three LAA regiments that had survived the post World War II demobilisation. One (the 25 LAA Regiment) had gone to Pakistan while two (the 26 and 27 LAA Regiments) were retained by India.

All three had continued to live up to the standards set during their formative years. That's history.

APPENDICES

APPENDIX 1

Japanese Air Forces and Air Tactics

Japan, although seemingly advanced in aerial tactics, entered the war with a narrow aerial doctrine, insufficient number of aircrafts and those of generally poor design (excluding the Mitsubishi A6M2 Zero, of course), too few aircrews and inadequate logistics for a war of attrition. Neither its army nor its naval air arm was prepared for the duration, violence or sophistication of the war to come.

The Japanese studied and trained hard at aerial tactics, but they failed to develop the airfield construction techniques and equipment, as well as the units, necessary to build air bases, maintenance, supply and dispersal facilities.

During the southern advance, the navy's 22nd Air Flotilla supported the attack into Malaya from three airfields in and around Saigon. Units were at full strength in aircraft and crews. Plentiful quantities of fuel and spare parts were available. The aircraft received excellent maintenance. 'Zeroes', for example, underwent a thorough overhaul every 150 hours of flight. As Japanese forces moved south, air units occupied, repaired and exploited captured enemy bases. Real problems developed, however, when those units reached undeveloped territories. Getting fuel, food and material to those bases determined whether the aircraft flew. Whether a base had been captured or built, it was nearly useless, however, if seaborne supplies could not reach it.

Mechanical complexity, battle damage and environmental stresses meant that maintenance was key to an aircraft's availability, its performance and whether the crew survived. Considering Japan's stressed economy, it should have been intolerable in terms of production and transportation to accept the loss of equipment that could have been repaired. Amazingly, the Japanese tolerated those losses.

Although a nucleus of well-trained army and navy maintenance men and armourers followed their aircraft south, maintenance units lagged behind during the early advances and were too few even when they caught up with the flying units. The army responded by sending forward individual maintenance units to plug gaps in maintenance coverage.

Although Japanese aircraft were generally lighter than Western counterparts and not so much in need of paved strips, occupying enemy airfields was never easy. Gasoline trucks were scarce and could be found at only a few of the large fields. Ground crews ordinarily had to refuel aircraft with hand pumps and barrels—a tedious process that slowed aircraft turnaround and consumed manpower.

The Japanese navy tables of organization and equipment specified that each air unit was to have extra aircraft in its organization, equal to one-third the operational complement. Yet by early April 1942, naval air units had no extras and were below their authorized operating strength.

Food at Japanese airfields was bad. Barracks were jungle slums. There were no laundry facilities, and men washed themselves in rivers, or under water-filled cans. Disease felled pilots and left serviceable aircraft grounded. Physical exhaustion lowered pilot performance, so that lesser-skilled opponents sometimes shot down veteran but feverish Japanese pilots.

Manpower became critical with no tractors, and ground crew wore themselves out pushing aircraft around fields. They worked at night to avoid Allied air attacks, only to fall victim to the malaria mosquito, which was most active at night. Men worked seven days a week in wretched weather at exhausting and mind-numbing tasks. Ground crew became nervous and irritable from lack of sleep. It took longer and longer to accomplish a given assignment. The number of minor as well as major accidents increased.

Japan's absence of standardization in weapons and equipment ranged from aircraft types to different engines, down to instruments and the smallest accessories. The army used a 24-volt electrical system, whereas the navy used a different voltage. Mounts to hold guns, cannons and rocket launchers varied between the two services. By the end of the war, Japan had produced at least 90 basic aircraft types (53 navy and 37 army) and 164 variations on basic types (112 navy and 52 army), making the logistician's jobs that much harder.

Mechanics at forward airfields were not trained well enough to correct many of the factory faults that were discovered when new aircraft arrived on station. The Japanese military also failed to master the supply, maintenance and medical problems that arose once its aerial units reached tropical zones far from their main depots.

Another limitation was that Home Island flight instructors were faced with too many students to train them effectively.

The Japanese had not experienced the logistical challenges that the Western powers had addressed during World War I and later relearned. Japan's politicians, generals and admirals completely misjudged the character and the duration of the war they launched in 1941. Poor aerial logistics planning, lack of foresight, a racist contempt for their enemies, a weak, shallow, narrow industrial base and an inability to appreciate supply requirements or to learn from their failures characterized their aviation effort throughout the entire war.

The army's air force, in particular, found itself poorly prepared to conduct the war it was eventually called on to fight. Having trained and equipped to fight a limited tactical war on land against the Soviet Union in the Far East, the Japanese Army Air Force (JAAF) lacked personnel capable of long-distance over-water navigation and possessed short-range aircraft that proved to be of limited use.

BOMBING TACTICS

Japanese bombers usually dropped their bombs at high altitudes, while flying in a V of V's formation. For protection, they had a tendency to depend more on altitude than on clouds. However, the enemy, quick to take advantage of bad weather, was likely to attack under very poor atmospheric conditions. These attacks were made in good formations, which were held after the bombs are dropped.

Japanese bomber formations usually consisted of 9, 18, or 27 planes.

In some sectors, Japanese Army Air Force bomber operations followed a fairly regular and characteristic pattern, roughly along these lines:

(1) Assembling units and moving them forward from rear airfields;
(2) Making photographic reconnaissance of the targets;
(3) Delivering the attack;
(4) Repeating the attack; and
(5) Withdrawing to rear airfields.

Bombing tactics called for close formations and straight, long runs, frequently at altitudes of 22,000 to 24,000 feet. Flights were often composed of 27 aircraft with 'Zero' fighters weaving above, below, or to port. When forced to bombardment ceiling by anti-aircraft, the Japanese used area bombing. The bombers flew in large formations and released bombs either in salvo or train, depending upon dispersion to encompass the target. Spot or precision bombing, which is more accurate than area bombing, was undertaken when there was no defensive anti-aircraft fire, and was delivered from altitudes below 10,000 feet.

Another variation was of the bomber formation consisting of a starboard flight of nine bombers at 21,000 feet, a middle flight of the same number in line but stepped up about 240 feet, with the right flight similarly stepped up, and the escorting aircraft weaving continuously. This method gave the starboard flight a greater field of fire and eliminated the need for fighter protection on this side.

Glide-bombing was usually undertaken by light bombers, the attack being launched at approximately 19,000 feet. When this manoeuvre was employed, aircraft nosed down toward the target at an angle of about 45 degrees and levelled off at from 3,000 to 2,500 feet before releasing the bombs. If the limits of anti-aircraft fire were clearly defined, swing-bombing was used. In this case planes flew toward the target until the perimeter of anti-aircraft fire was reached, and, making a banked turn, released their bombs with a centrifugal 'throw.' The inaccuracy of this method is obvious.

In dive-bombing, the practice of the Japanese was to approach the objective at about 12,000 feet and gradually work down to about 7,000 feet. As the attack developed, the aircraft went into dives of about 60 to 80 degrees with diving flaps lowered. The dives culminated in pull-outs ranging down to 800 feet. As many as 60 to 70 Japanese aircraft were used in these dive-bombing attacks. Japanese escort fighters had a tendency to work in pairs.

In one instance, 21 Japanese medium bombers, accompanied by a large fighter escort, made a high-level attack from 23,000 feet, while a smaller formation of light bombers carried out a low-level bombing and strafing attack.

The medium bombers pressed home their attack despite the fact that a large percentage of them were destroyed or damaged by Allied fighters before reaching the target. The original formation was broken, but the bombers were still able to reach their objective when reformed into three flights, each consisting of four bombers and flying in a tight diamond bow formation. The Japanese apparently had little fire control, and the bombers carried out no evasive tactics except to nose down after passing the bomb release line. The escorting fighters appeared to use a generous amount of white tracer.

The light bombers flew at approximately 200 to 220 miles per hour while making their bombing and strafing runs.

GROUND ATTACK

Tactical practices employed by Japanese fighter aircraft varied according to the situation and the type of opposing aircraft encountered. A formation frequently used by the Japanese for ground attack was a 'V' of three aircraft flanked by echelons of two. When fighter opposition is encountered the echelon 'pairs' meet it, while the 'V' executes the ground attack.

In early operations, Japanese employed two principal methods of ground attack, which are still considered effective. In the first attack, fighters come in just over the trees, dive on an Allied airdrome, machine-gun grounded aircraft or anti-aircraft emplacements and then fly away in horizontal formation at low altitude. In the case of one such attack, a fighter remained to circle the field at an altitude of 12,000 to 18,000 feet, apparently to observe the results of the attack. Shortly afterwards, the Japanese launched a new attack, probably making use of information gained from the observation.

In the second method of attack, a Japanese fighter flew over the airfield and attacked with machine guns while one or more flights of aircraft remained at altitudes of 16,000 to 20,000 feet. When defending Allied fighters attempted to get into the air, the Japanese planes immediately dove upon them at high speed. In a similar attack, the Japanese fired tracer followed by 20 mm. explosive. They passed the targets at a height scarcely over 25 feet, flew about 50 yards beyond the edge of the field, and after making easy turns, repeated the attack again and again.

APPENDIX 2

Anti-Aircraft Guns

Indian anti-aircraft regiments and batteries manned a variety of AA guns, ranging from 18 pounder Mk I and 20-mm. Hispano guns to 3.7-in. Heavy AA and 4-in. naval AA guns. The guns operated by the Indian AA gunners are given out in succeeding paragraphs.

20-mm. Hispano AA Gun

HS.404 was an autocannon originally designed and produced by Spanish/French company Hispano-Suiza in the mid-1930s and was widely used as an aircraft, naval and land-based weapon by French, British, American and other military services. The versions based on British development were known as 20-mm. Hispano.

18-Pounder Mk I

The Ordnance QF 18-pounder, or simply the 18-pounder gun, was the standard British Empire field gun of the First World War era. In early 1915, a number of 18-pounder guns were mounted on pedestals, with the addition of a second recuperator and retaining catch for the cartridge case at high angle, in an attempt to come up with a workable anti-aircraft gun. The gun's relatively low muzzle velocity, and the unsatisfactory ballistic characteristics of its shrapnel shell at high angles, made it a marginal performer in such a role.

40-mm. Bofors LAA Gun

The Bofors 40-mm. gun, often referred to simply as the Bofors gun was an anti-aircraft autocannon designed in the 1930s by the Swedish arms manufacturer A.B. Bofors. The British Army had first examined the 'QF 40-mm. Mark I' in 1937. For aiming, the gun used a complex mechanical analogue computer, the Kerrison Predictor, which drove the laying electrically and was operated by a three-man crew. The version with the Kerrison Predictor came to be called the 'QF 40-mm. Mark III'. As the Kerrison was difficult to set up for use in combat, a new sight, called the 'Stiffkey Sight' was developed that made the gun easier to use. The other versions included the much lighter two-wheeled carriage for airborne use and a self-propelled version with the gun mounted on the chassis of a Morris Commercial four-wheel drive lorry.

QF 3-inch 20 cwt

The QF 3-inch 20 cwt anti-aircraft gun was the standard anti-aircraft gun used in the home defence of the United Kingdom against German airships and bombers and on the Western Front in World War I and was itself based on Vickers naval 3-inch (76 mm.) QF gun. During World War II most of these guns were in the heavy anti-aircraft (HAA) role until replaced by the new 3.7-inch gun but some were deployed as light anti-aircraft guns (LAA) for airfield defence, primarily by the RAF regiment when it was formed in 1942, until replaced by 40-mm. Bofors guns.

QF 3.7-inch AA gun

The QF 3.7-inch AA was the primary heavy anti-aircraft gun during World War II. The gun entered service just before the war and was used throughout in all theatres except the Eastern Front. It was produced in two versions—mobile and static. The mobile version had a wheeled carriage with four folding outrigger trails and levelling jacks. The wheels were lifted off the ground or removed when the gun was brought into action. The other used a travelling platform with detachable wheels for guns to be used in static positions but which could be re-positioned.

Gun Characteristics

Gun	*Calibre*	*Muzzle velocity*	*Weight of Shell/Round*	*Rate of Fire*	*Effective Range*	*Remarks*
20-mm. Hispano	20 mm.	870 m/s		600-700 rpm	6,500 m.	Gas operated
18-Pounder	83.8 mm.	492 m/s	8.4 kg	20 rpm	6,000 m.	
40-mm. Bofors	40 mm.	881 m/s	0.9 kg	120 rpm	7,160 m.	
QF 3-in. 20 cwt	76.2 mm.	760 m/s	5.7 kg	16-18 rpm	4,900 m.	
QF 3.7-inch AA Gun	94 mm.	814 m/s	13 kg	10-20 rpm	5,600 m.	9,000 m. effective ceiling

Source: Auhtor's notes.

APPENDIX 3

Light AA/Anti-Tank Regiments

Four Indian anti-tank regiments converted or partially converted to the LAA/A Tk regiments during World War II but they were never employed in their new role excepting the 2nd LAA/A Tk Regiment that served with the 23rd Division for a brief period in 1943-4. These regiments were:

1st LAA/A Tk Regiment

The regiment was raised at Hyderabad as the 'K' Anti-Tank Regiment on 1 August 1940 and was redesignated as the 1st Anti-Tank Regiment on 2 April 1941. The regiment served in Iraq in 1941 with the 8th and 10th Indian Infantry Divisions, returning to India in March 1943. It disembarked at Karachi on 13 March 1943 before moving to quarters at Sialkot two days later. The regiment was notified on 15 August 1943 that it had been earmarked for conversion as a LAA/A Tk regiment. Whilst some preparatory training was undertaken, no guns had been received till December 1943. It converted to become an anti-tank/ mortar regiment in 1944, with three batteries, each equipped with twelve 6-pounder anti-tank guns and twelve 3-inch mortars. By October 1944, the regiment had converted to this new organisation.

2nd LAA/A Tk Regiment

The regiment was raised at Hyderabad as the 'L' Anti-Tank Regiment on 1 April 1941, changing its designation to the 2nd Anti-Tank Regiment the next day. The regimented batteries were the 5th, 6th, 7th and 8th Anti-Tank Batteries. In February 1942, the regiment arrived in Burma and served under the command of the 17th Indian Infantry Division. By early 1943, after its return from Burma, the regiment had become divisional troops to the 23rd Indian Infantry Division.

During July 1943, the regiment was notified that the 5th and 8th

Batteries were to convert as the light anti-aircraft batteries and were to be equipped with 40-mm. Bofors guns and later with 20-mm. Hispano guns for both anti-aircraft and ground roles.

On 22 July 1943, the regiment began preparing for the conversion, liaising with the 28th LAA Regiment, RA, which not only conducted a familiarisation cadre but also provided Bofors guns for gun drill demonstration. The regiment was, meanwhile, informed that it would be equipped with 'Oerlikon' 20-mm. guns. As the AA guns were awaited, both batteries continued to train with the 2-pounder anti-tank gun for their original roles. In November 1943, the regiment was told to expect delivery of four Hispano guns for use in training by the 5th and 8th Batteries. This was followed by orders that the AA guns for the two batteries would be the 40-mm. Bofors after all, two of which arrived in February 1944. With this, at least, the gun drill training began. However, the 5th and 8th Batteries continued to retain their 2-pounder anti-tank guns for the remainder of the 'campaigning season.' In March 1944, detachments of the regiment began to prepare defensive positions at the 'Thoubal Box' in anticipation of the Japanese offensive. Against this background of imminent action, on 16 March 1944, the 8th Battery exchanged its 2-pounder anti-tank guns for twelve 40-mm. Bofors guns. The battery was expected to have completed conversion by 1 April 1944. The regiment, together with the 6th, 7th and 8th Batteries was part of the defences at the Thoubal Box, the 5th Battery having been attached to the 123rd Indian Infantry Brigade and sent to Imphal.

The regiment officially converted to become the 2nd Anti-Aircraft/Anti-Tank Regiment, IA, on 1 April 1944. However, during the month it was decided that the 5th Battery would not convert to become an anti-aircraft battery until the 8th Battery had been fully trained. As the 8th LAA Battery was already in existence (as part of 2nd LAA Regiment), the 8th Battery was renumbered as the 208th LAA Battery.

Meanwhile, inside the Thoubal Box, the 208th Battery had three Bofors guns in action. The battery moved to Bull Box (Palel) during May and June 1944, accompanied by RHQ and rear parties. The location statement for the divisional artillery of the 23rd Indian Infantry Division for 28 June 1944 gives the disposition of the regiment as follows:

HQ, 2nd AA/A Tk Regiment	Bull Box
5th Anti-Tank Battery	Bull Box
6th Anti-Tank Battery	Langgol
7th Anti-Tank Battery	Patiala Ridge
208th LAA Battery	Langgol, less one troop at Palel

At the end of July 1944, the regiment left Bull Box for the Shenam area where the main responsibility was to patrol the area in infantry role, on the lookout for Japanese stragglers. The regiment returned to Bull Box on 7 August 1944 and the following day handed in all its remaining 2-pounder anti-tank guns. On 9 August 1944, the 208th Battery set off for Manipur Road where it handed in its Bofors guns. The RHQ with the 5th, 6th and 7th Anti-Tank Batteries moved to Manipur Road on 10 August 1944 as the first step in the journey to Shillong, where it arrived on 13 August 1944.

Indian divisional anti-tank regiments were once again the subject of reorganisation from August 1944, when they were reorganised to consist of three batteries, each equipped with twelve 6-pounder anti-tank guns and twelve 3-inch mortars. The 2nd Regiment converted to become one such anti-tank/mortar unit during August 1944, drawing its first 3-inch mortars at Shillong on 21 August 1944.

During August 1944, the 208th was disbanded at Shillong as part of the reorganisation of the regiment to become an anti-tank/mortar regiment, bringing an end to the only Indian LAA/A Tk regiment.

7th LAA/A Tk Regiment

The 7th Anti-Tank Regiment was raised at Kumbhargaon on 1 May 1942 with the 25th , 26th and 27th Batteries. From early 1943 to March 1944, the regiment served with the 25th Indian Infantry. The regiment moved to Madras where by April it began converting to become an AA/A Tk regiment. Probably in August 1944, it was reorganised as an anti-tank/mortar regiment. By the time it left Madras in late 1944 for the Arakan to rejoin the 25th Indian Infantry Division, it was composed of the 9th, 25th and 26th Anti-Tank Batteries, equipped with both 6-pounder anti-tank guns and 3-inch mortars.

15th (Punjab) LAA/Anti-Tank Regiment

The 15th (Punjab) Anti-Tank Regiment was raised on 1 August 1942, by the conversion of the Machine Gun Battalion, 15th Punjab Regiment. It formed part of General Reserve of GHQ India until 19 July 1943 when it became part of IV Corps. In September 1943, the regiment moved from Babina, near Jhansi, and came under the command of 151 Lines of Communication Sub-Area for administration. At this time, the regiment was organised as a four battery regiment, composed of the 28th, 29th,

30th and 31st Batteries. The regiment remained listed as Corps Troops, IV Corps, until October 1944 when it came under command of XXXIII Corps. The regiment converted to LAA/A Tk Regiment with three batteries—two LAA and one A Tk. These were numbered as the 203rd and 204th AA Batteries and the 31st A Tk Battery. It was disbanded on 7 November 1944 while located at Imphal.

APPENIDX 4

Raising of Indian Air Defence Units

Unit/Establishment	*Date of Raising*	*Place of Raising*	*Remarks*
Technical Training Battery	15 September 1940	Karachi	Upgraded and redesignated as AATC on 1 January 1941.
1 Indian Heavy AA Regiment	1 October 1940	Bombay	Ceased to exist in Singapore.
AATC	1 January 1941	Karachi	Technical Training Battery upgraded and redesignated as 1 AATC.
1 Indian LAA Regiment	1 March 1941	Karachi	2, 3 and 6 LAA Batteries, redesignated as 25 LAA Regiment w.e.f. 25 January 1947. 2 LAA Battery replaced by 7 LAA Battery of 2 LAA Regiment in 1946 making it all-Punjabi Mussalman Regiment. Transferred to Pakistan on 15 August 1947 (with 3,6,7 Btys). Renumbered as 6 LAA Regiment, RPA.

2 Heavy AA Regiment	1 April 1941	Karachi	1, 7 and 8 HAA Batteries. Redesignated as 18 HAA Regiment w.e.f. 25 January 1947. 8 HAA Battery replaced by 19 HAA Battery from 7 Rajput HAA Regiment in 1946. Transferred to Pakistan 15 August 1947 with 1, 7, 19 Batteries. Renumbered as 5 HAA Regiment, RPA.
3 Heavy AA Regiment	1 June 1941	Karachi	5, 6 and 11 HAA Batteries. Converted to 2 Medium Regiment August 1945.
2 LAA Regiment	1 October 1941	Drigh Road	HQ & 7-9 LAA Btys. Renumbered as 26 LAA Regiment, RIA. Survived Partition. One of the two AA Regiments as part of Indian Artillery.
4 Heavy AA Regiment	1 November 1941	Drigh Road	HQ & 9-11 HAA Btys. Reorganised on 15 Sep. 1942 to HQ & 9 Ind Hy AA Bty and 11 Ind Lt AA Bty (10 & 11) Hy AA. Btys ceased to belong to it w.e.f. 15 Nov. 1942. The regiment ceased to administer 9 Hy and 11 Lt Btys as well and Regimental HQ was redesignated as HQ 11 Ind Hy AA Regiment on 15 Jan. 1943. Demobilised.
No 2 AATC	1 January 1942	Deolali	

3 LAA Regiment	1 February 1942	Malir Cantt	The original composition was HQ & 10-12 LAA Btys. 10 and 11 LAA Btys ceased to belong to this regiment w.e.f. 15 Sep. 1942 and were replaced by 13 and 14 LAA Btys. Renumbered as 27 LAA Regiment, RIA, w.e.f. 25 Jan. 1947. Survived Partition. One of the two AA regiments as part of Indian Artillery.
7 Rajput HAA Regiment	1 February 1942	Majerhat	HQ & 18-20 Hy AA Btys. Converted from 8/7 RAJPUT.
6 Punjab LAA Regiment	15 February 1942	Waldinadi	HQ & 19-21 LAA Btys. Converted from 8/1 Punjab Regiment. Converted 6 Punjab Para LAA Regiment. Renumbered 28 Para LAA Regiment August 1945.
HQ 1 AA Brigade	20 February 1942	Calcutta	
5 HAA Regiment	1 March 1942	Malir Cantt	HQ & 12-14 HAA Batteries. Converted to 4 Medium Regiment August 1945.
No. 1 Gun Operations Room (GOR)	1 April 1942	Calcutta	Redesignated as No. 1 AAOR w.e.f. 12 Nov. 1942.
8 Rajput HAA Regiment	1 April 1942	Ahmed-abad	HQ & 21-23 HAA Btys.

7 Sikh LAA Regiment	1 April 1942	Peshawar	HQ & 22-24 LAA Btys. Converted from 9/11 Sikh Regiment. The regiment (less 22 SIKH LAA Bty) disbanded w.e.f. 30 Sep 1942.
8 Sikh LAA Regiment	1 April 1942	Nowshera	HQ & 25-27 LAA Btys. Converted from 8/11 Sikh Regiment.
9 Rajputana LAA Regiment	1 May 1942	Mhow	HQ & 28-30 LAA Btys. Converted from 9/6th Rajputana Rifles. 28 LAA Bty of the regiment disbanded w.e.f. 30 Sep. 1943 and was replaced by 22 Sikh LAA Bty.
2 GOR	1 May 1942	Digboi	Redesignated as No. 2 AAOR w.e.f. 12 Nov. 42.
3 GOR	1 May 1942	Tatanagar	Redesignated as No. 3 AAOR w.e.f. 12 Nov. 42.
20 HAA Battery	1 May 1942	Calcutta	Forms part of 7 (Rajput) HAA Regiment.
HQ 2 AA Brigade	15 May 1942	Calcutta	
9 Punjab HAA Regiment	1 June 1942	Lahore	HQ & 24-26 HAA Btys. Converted to 12 Field Regiment.
HQ 3 AA Brigade	5 June 1942	Bombay	
3 AATC	1 July 1942	Deolali	Shifted to Mehgaon 12 July 1943.
13 LAA Battery	15 August 1942	Deolali	Forms a part of 3 Ind LAA Regiment w.e.f. 15 Sep. 1942.
14 LAA Battery	15 August 1942	Deolali	Forms a part of 3 Ind LAA Regiment w.e.f. 15 Sep. 1942.
6 HAA Regiment	1 September 1942	Deolali	HQ & 15-17 HAA Btys. Renumbered as 19 HAA Regiment w.e.f. 25 January 1947.

4 LAA Regiment	1 November 1942	Mehgaon	HQ & 15-17 LAA Btys. Renumbered as 29 LAA Regiment, RIA. w.e.f. 25 Jan. 1947.
23 HAA Battery	1 September 1942	Bangalore	Forms part of 8 (Rajput) HAA Regiment.
5 LAA Regiment	15 December 1942	Mehgaon	HQ & 18, 31, 32 LAA Btys.
10 HAA Regiment	15 November 1942	Mehgaon	HQ & 27-29 HAA Btys. Converted to 14 Field Regiment August 1945.
No. 4 AATC	1 February 1943	Calcutta	For WAC(I).
10 LAA Regiment	1 February 1943	Mehgaon	HQ & 33-35 LAA Btys.
11 HAA Regiment	15 January1943	Mehgaon	HQ 4 HAA Regiment designated as RHQ 11 HAA Regiment w.e.f. 15 January 1943. HQ & 30-32 HAA Btys. Converted to 15 Field Regiment August 1945.
11 LAA Regiment	1 April 1943	Mehgaon	HQ & 36-38 LAA Btys.
17 LAA Regiment	1 April 1943	Malir	Regimental HQ only
101 LAA Bty	1 February 1943	Malir	
102 LAA Bty	1 April 1943	Malir	Reorganised and redesignated as 102 (Comp) AA Bty with effect from 1 July 1943.
103 LAA Bty	1 May 1943	Malir	
20 HAA Regiment	1 April 1943	Malir	Converted to 9 Anti-Tank Regiment August 1945.
501 HAA Battery	1 February 1943	Malir	

502 HAA Battery	1 April 1943	Malir	Reorganised and redesignated as 502 (Comp) AA Battery with effect from 1 July 1943.
503 HAA Battery	1 May 1943	Malir	Converted to 3 Anti-Tank Regiment August 1945.
12 HAA Regiment	1 May 1943	Mehgaon	HQ & 33 to 35 HAA Btys.
25 HAA Regiment	1 May 1943	Malir/Basra	
12 LAA Regiment	15 May 1943	Mehgaon	HQ & 39-41 LAA Btys.
13 HAA Regiment	1 June 1943	Mehgaon	HQ & 36 to 38 HAA Btys. Converted to 13 Field Regiment August 1945.
18 LAA Regiment	1 June 1943	Malir	Converted to 11 Field Regiment August 1945.
104 Lt AA Battery	1 June 1943	Malir	
105 Lt AA Battery	1 August 1943	Malir	
21 HAA Regiment	1 June 1943	Malir	Converted to 16 Field Regiment August 1945.
504 HAA Battery	1 June 1943	Malir	
505 HAA Battery	1 August 1943	Malir	
506 HAA Battery	1 September 1943	Malir	
13 LAA Regiment	1 August 1943	Mehgaon	HQ & 42-44 LAA Btys.
Reserve Mobile AA Regiment	15 August 1943	Chindwara	
14 HAA Regiment	1 September 1943	Mehgaon	HQ & 39 to 41 Hy AA Btys. Converted to 3 Medium Regiment August 1945.

15 HAA Regiment	1 October 1943	Malir	Consisting of HQ and 42 HAA Bty on date of raising. 43 and 44 HAA battery later affiliated with it. Reorganised as 15 LAA Regt w.e.f. 1 June 1944.
42 HAA Battery	1 October 1943	Malir	Reorganised as 42 LAA Battery 1 June 1944.
43 HAA Battery	1 November 1943	Malir	Reorganised as 43 LAA Battery 1 June 1944.
44 HAA Battery	1 December 1943	Malir	Reorganised as 44 LAA Battery 1 June 1944.
14 LAA Regiment	15 October 1943	Mehgaon	HQ & 45, 46, 47 LAA Btys.
16 HAA Regiment	7 January 1944	Mehgaon	Converted as 16 LAA 1 June 1944 with RHQ & 106-108 LAA Batteries.
1 AA Ops Room	22 February 1946		Disbanded before Partition.
2 AA Ops Room	22 February 1946		Disbanded before Partition.
HQ 11 Army Group RIA (AA)	28 February 1946	Gulunche	
HQ 12 Army Group RIA (AA)	28 February 1946	Uruli	

Source: Statistical Review of Personnel, Indian Army, vol. I to IV: 1939-1948, compiled by Army Statistical Organisation (A.S.O./2), published by Manager, Government of India Press, New Delhi, 1948.

APPENDIX 5

Composition of Indian AA Regiments 1940-1945

Regiment/ Battery	*Class*
HAA Regiments	
1 HAA Regiment	Punjabi Mussalman (4th HAA Battery of Punjabi Brahmins and Sikhs)
2 HAA Regiment	
1 HAA Battery	Punjabi Mussalman
7 HAA Battery	Punjabi Mussalman
8 HAA Battery	Ahir
3 HAA Regiment	
5 HAA Battery	Punjabi Mussalman
6 HAA Battery	Madrassi
11 HAA Battery	Punjabi Brahmin
5 HAA Regiment	
12 HAA Battery	Punjabi Mussalman
13 HAA Battery	Madrassi
14 HAA Battery	Ahir and Madrassi
6 HAA Regiment	Madrassi
7 (Rajput) HAA Regiment	
18 HAA Battery	Rajput
19 HAA battery	Punjabi Mussalman
20 HAA Battery	Rajput and Punjabi Mussalman
8 (Rajput) HAA Regiment	Rajput, Punjabi Mussalman (exact details not known)
9 (Punjab) HAA Regiment	
10 HAA Battery	Madrassi
24 HAA Battery	Punjabi Mussalman
26 HAA Battery	Gujjar and Ahir

10 HAA Regiment	Madrassi
11 HAA Regiment	Madrassi
12 HAA Regiment	Madrassi
13 HAA Regiment	Madrassi
14 HAA Regiment	Madrassi
15 HAA Regiment	Madrassi
20 HAA Regiment	Madrassi
21 HAA Regiment	Madrassi
25 HAA Regiment	Madrassi
LAA Regiments	
1 LAA Regiment	
2 LAA Battery	Punjabi Brahmin
3 LAA Battery	Punjabi Mussalman
6 LAA Battery	Punjabi Mussalman
2 LAA Regiment	
7 LAA Battery	Punjabi Mussalman
8 LAA Battery	Ahir
9 LAA Battery	Madrassi
3 LAA Regiment	Madrassi
4 LAA Regiment	Madrassi
5 LAA Regiment	Madrassi
6 (Punjab) LAA Regiment	
19 LAA Battery	Punjabi Mussalman
20 LAA Battery	Sikh
21 LAA Battery	Rajput
7 (Sikh) LAA Regiment	
23 LAA Battery	Sikh
25 LAA Battery	Punjabi Mussalman
27 LAA Battery	Sikh
8 (Sikh) LAA Regiment	Sikhs, Punjabi Mussalman (exact details not known)
9 (Rajputana Rifles) LAA Regiment	
10 LAA Battery	Ahir
29 LAA Battery	Rajput
30 LAA Battery	Punjabi Mussalman

10 LAA Regiment	Madrassi
11 LAA Regiment	Madrassi
12 LAA Regiment	Madrassi
13 LAA Regiment	Not known
14 LAA Regiment	Not known
17 LAA Regiment	Madrassi
18 LAA Regiment	Madrassi

Source: History of Army Air Defence, AAD Association, New Delhi, 2008, p. 26.

APPENDIX 6

Indian AA Order of Battle

AA Brigades—1st, 2nd, 3rd
HAA Regiments—1st to 16th, 20th, 21st, 25th
LAA Regiments—1st to 18th
Composite Batteries—102nd and 103rd AA Battery, 502 AA Battery

AA Order of Battle—Eastern and Southern Commands India 1943-5

April 1943

AA Brigade	*Deployment Area*	*Regiments and Batteries*	
		Royal Artillery	*Indian Artillery*
9th AA Brigade, RA	Digboi, Dimapur, Lumding, Tinsukhia, Assam airfields	66, 67 (less Bty) HAA Regts	2 Indian, 5 Indian HAA Regts
		69 LAA Regt, 143/78 LAA Bty	1 Indian LAA Regt
13th AA Brigade, RA	Chittagong, Agartala, Bengal and Assam airfields	8 HAA Regt, 8, 202/56 HAA Btys	14/5 HAA Bty
		77 LAA Regt, RA	8 Indian LAA Regt, 1, 4 Indian LAA Btys
1st Indian AA Brigade	Calcutta, Alipore, Dum Dum, Chittagong	95 HAA Regt, 216/70 HAA Bty	7 Indian HAA Regt
		59 LAA Regt,	9 Indian LAA Regt, 13/3 Indian LAA Bty
2nd Indian AA Brigade	Calcutta, Jamshedpur, Asansol, Pandeshwar	53 HAA Regt	3 Indian HAA Regt (less Bty), 11/4 Indian HAA Bty
		44 LAA Regt	

3rd Indian AA Brigade	Bombay, Madras,	56, 76 (less Bty) HAA Regts	-
	Vizagapatam	33, 118 LAA Regts	6 Indian, 7 Indian LAA Regts

4 Corps: 187/67 HAA Bty, 78 LAA Regt (less Bty)
15 Corps: 36 LAA Regt
Divisional Regts: 28, 60 LAA Regts, 24, 82 AA/A-Tk Regts

September 1943

AA Brigade	*Deployment Area*	*Regiments and Batteries*	
		Royal Artillery	*Indian Artillery*
9th AA Brigade, RA	Digboi, Dimapur,	66, 67 (less Bty) HAA Regts	2 Indian, 5 Indian HAA Regts
	Lumding, Tinsukhia, Assam airfields, Ledo	69, 118, LAA Regts, 143/78 LAA Bty	5 Indian LAA Regts, 4 Indian LAA Bty
13th AA Brigade, RA	Chittagong, Agartala, Bengal	8, 70 HAA Regts, 8, 202/56 HAA Btys	6 Indian HAA Regt
	and Assam airfields, Cox's Bazaar	36, 77 LAA Regt, RA	3, 8 Indian LAA Regts, 10 Indian LAA Bty
1st Indian AA Brigade	Calcutta, Bengal	53, 95 HAA Regts	11 Indian HAA Regt
		-	4, 9 Indian LAA Regts
2nd Indian AA Brigade	Calcutta, West Bengal	-	3 , 4 Indian HAA Regts
	airfields and depots	44 LAA Regt	–
3rd Indian AA Brigade	Bombay, Madras,	56 (less bty) HAA Regt	8 Indian HAA Regt
	Trichinoply	33 LAA Regt	6 Indian, 7 Indian LAA Regts
14th WA AA Brigade	Bombay	1,2,3 HAA Regts (WA)	–

March 1944

Fourteenth Army		
4 Corps	HAA	67 HAA Regt
	LAA	28, 78 LAA Regts, 104/18 Indian LAA Bty
15 Corps	HAA	8 HAA Regt, 8 HAA Bty
	LAA	36 LAA Regt
33 Corps	LAA	59, 60 LAA Regts
Div Regts	AA/A Tk	24, 55, 56, 82, 100, 122 AA/A Tk Regt, 1 WA AA/A Tk Regt
13 AA Bde*	HAA	70 HAA Regt, 6, 7 Indian HAA Regts, 35/12 Indian HAA Bty
	LAA	77 LAA Regt, 3, 6 & 8 Indian LAA Regts, 10 Indian LAA Bty
14(WA) AA Bde**		1, 2, 3 WA HAA Regts 118 LAA Regt, 2 & 8 (Less Bty) Indian Regts, 10 Indian LAA Bty

Notes: * In support of 15 Corps in Chittagong, East Bengal and Bawli.
** In Assam and Dimapur Base.

AA Brigade	*Deployment Area*	*Regiments and Batteries*	
		Royal Artillery	*Indian Artillery*
9th AA Brigade, RA	Bombay, Madras, Vizagapatam, Trichinoploy	56, 101 HAA Regts 33 LAA Regt	8 Indian Regt, 10 Indian HAA Bty –
1st Indian AA Brigade	Calcutta, Alipore, Asansol	95 HAA Regt -	11 Indian HAA Regt 4, 9 Indian LAA Regts, 13/3 LAA Bty
2nd Indian AA Brigade	Calcutta, Jamshedpur, Tatanagar, Bengal airfields	53 HAA Regt, 44 LAA Regt	3 Indian HAA Regt, 11/4 HAA Bty –

3rd Indian AA Bde	Digboi, Ledo, Assam airfields, Tinsukhia	66 HAA Regt	2, 5 Indian HAA Regts, 19/7 HAA Bty
		69 LAA Regt (less dets)	1 Indian (less Bty), 5 Indian LAA Regts

December 1944

Fourteenth Army		
4 Corps	HAA	67 HAA Regt
	LAA	207/69 LAA Bty, 28 Regt in Reserve (Ranchi)
15 Corps	HAA	8 HAA Regt
	LAA	36 LAA Regt, 27/8 Indian LAA Bty
33 Corps	HAA	52, 101 HAA Regts
	LAA	44 LAA Regt, Div Regts (AA/A Tk Regts)
Div Regts	AA/A Tk	24, 55, 56, 82, 100, 122 AA/A Tk Regt, 1 WA AA/A Tk Regt
13 AA Bde*	HAA	6 Indian, 7 Indian HAA Regts, 15 East African HAA Regt
	LAA	3, 6 & 11 Indian LAA Regts, 25/8 Indian and 1 WA LAA Bty
14(WA) AA Bde**		1, 2, 3 WA HAA Regts
	LAA	69 LAA Regt (less Bty), 2 & 4 (Less Bty), 18 Indian Regts

Notes: * In support of 15 Corps in Chittagong, East Bengal and Arakan (Bawli and Maungdaw).
** In Imphal Dimapur, Surma Valley, Sylhet.

AA Brigade	*Deployment Area*	*Regiments and Batteries*	
		Royal Artillery	*Indian Artillery*
9th AA Brigade, RA	Bombay, Poona, Madras	–	2 Indian Regt, 10 Indian HAA Bty
		56 LAA Regt (in training)	8 (less Bty) and 9 Indian LAA Regt

1st Indian AA Brigade	Calcutta	–	3 Indian HAA Regt
3rd Indian AA Bde	Digboi, Ledo, Assam airfields	66 HAA Regt, 308/2 WA HAA Bty	5 Indian HAA Regt
		207/69 LAA Bty	1, 5 Indian LAA Regts

The 3rd Indian and 9 AA Brigades exchanged locations in March/April 1944.

The LAA Batteries of the six LAA/A Tk Regiments were broken up in the second half of 1944. Other LAA Regts were under corps command.

March 1945

Fourteenth Army		
4 Corps	HAA	52, 67 HAA Regts
	LAA	28 LAA Regt, 1 Indian LAA Regt, 32/5 Indian LAA Bty
15 Corps	HAA	8 HAA Regt, 155/15 WA HAA Bty
	LAA	36 LAA Regt
33 Corps	HAA	101 HAA Regt, 19/7 HAA Bty
	LAA	44 LAA Regt, 2 Indian LAA Regt, 239/69 LAA Bty
13 AA Bde*	HAA	6 Indian, 25 Indian HAA Regts, 15 EA, 4 WA HAA Regts
	LAA	3, 11 & 15 Indian LAA Regts, 15/4 Indian LAA Bty
14 (WA) AA Bde**		1, 2, 3 WA HAA Regts
	LAA	69 LAA Regt (less Bty), 4, 18 Indian LAA Regts
3rd Indian AA Bde***		5, 15 Indian HAA Regts, 66 HAA Regt
	LAA	5 Indian LAA Regt

Notes: * In Chittagong, Feni, Comilla, Dohazari, Bawli, Cox's Bazar
** In Imphal, Dimapur, Surma Valley, Sylhet
*** In Digboi, Ledo, Assam

AA Brigade	*Deployment Area*	*Regiments and Batteries*	
		Royal Artillery	*Indian Artillery*
9th AA Brigade, RA	Bombay (in training)	–	2, 7, 8 Indian Regts
		–	8 and 9 Indian LAA Regt
1st Indian AA Brigade	Calcutta	–	3 Indian HAA Regt
24th AA Brigade	Jorhat, awaiting move	207/69 LAA Bty	6 Indian LAA Regt

May 1945

Fourteenth Army		
4 Corps	LAA	28 LAA Regt, 32/5 Indian LAA Bty
15 Corps	HAA	8 HAA Regt
	LAA	36 LAA Regt (awaiting withdrawal)
33 Corps	HAA	101 HAA Regt
	LAA	44 LAA Regt
13 AA Bde*	HAA	6 Indian, 25 Indian HAA Regts, 15 EA HAA Regts
	LAA	3, 4, 11 & 15 Indian LAA Regts
24 AA Bde**	52 HAA Regt, 5 Indian HAA Regt	
	LAA	1, 2, 5 Indian LAA Regts

Notes: * In Chittagong, Akyab Ramree
** 61 Agra (AA) At Rangoon, Meiktila, Toungoo

India	
3rd Indian AA Bde	HQ only, awaiting move to Rangoon
9th AA Bde (60AGRA/AA)	2, 7, 8 Indian HAA Regts, Bty HKSRA
14 WA AA Bde	1, 2, 3, 4 WA HAA Regts
Non operational	69 LAA Regt

Notes: 3rd Indian AA Bde relieved 2nd Indian AA Bde in June, taking over regiments
1st Indian AA Bde withdrawn from Calcutta but dates not established

Source: The details have been taken from War Diaries of AA Brigades and Regiments as also the R.N. Routledge, *History of the Royal Regiment of Artillery: Anti-Aircraft Artillery, 1914-1955*, Brassey's, London, 1994.

APPENDIX 7

Honours and Awards

S No	*Army No*	*Rank*	*Name*	*Unit*	*Theatre*	*London Gazette Notification dated*
Distinguished Service Order (DSO)						
1	8953	Lt Col	John Rowley Williamson	1 Indian HAA Regiment	Malaya	1 August 1946
Military Cross (MC)						
1	162868	Lt	John Ashcroft Hopson	1 Indian HAA Regiment	Malaya	13 December 1945
2	201670	Lt	F. D. Webber	3 Indian LAA Battery	Burma	9 March 1943
3	201657	Lt	L.L . Nash	3 Indian LAA Battery	Burma	9 March 1943

Member of the Most Excellent Order of the British Empire						
1	3957	Lt	Sultan Khan	1 Indian LAA Regiment		N.K.
Indian Order of Merit (2nd Class)						
1	7586	Subedar	Nur Khan	3 Indian LAA Battery	Burma	9 March 1943
Order of British India (2nd Class)						
1		Subedar Major	Jewan Khan	2 Indian LAA Regiment	N.K.	N.K.
Indian Distinguished Service Medal (IDSM)						
1	AAA/1165	Havildar	Sham Lal	1 Indian HAA Regiment	Malaya	N.K.
2	AAA/5122	Gunner	Balbir Singh	1 Indian HAA Regiment	Malaya	N.K.
3	AAA/1991	Havildar	Hakim Khan	3 Indian LAA Battery	Burma	9 March 1943
4	AAA/2665	Dvr Mech	Rehmat Ullah	3 Indian LAA Battery	Burma	9 March 1943
5	AAA/4062	Gunner	Suda Rang	3 Indian LAA Battery	Burma	9 March 1943
6	AAA/8837	Gunner	Hari Singh	8 Sikh LAA Battery	Burma	16 March 1944

7	AAA/7510	Gunner	Mehnga Singh	8 Sikh LAA Battery	Burma	16 March 1944
8	AAA/5562	Havildar	Surat Singh	2 LAA/AT Regiment	Burma	16 November 1944
Mention in Despatches						
1	N.K.	Lt	P. I. Taylor	3 Indian LAA Battery		N.K.
2	1560427	Sgt	A Clough	3 Indian LAA Battery		N.K.
3	17777	Jamadar	Abit Ali	3 Indian LAA Battery		9 March 1943
4	AAA/2016	Havildar	Sardar Khan	3 Indian LAA Battery		9 March 1943
5	AAA/1874	L/Naik	Sher Khan	3 Indian LAA Battery		9 March 1943
6	AAA/1858	Havildar	Abdul Ghani	N.K.		17 Sep. 1946
7	AAA/2193	Havildar	Alam Khan	N.K.		17 Sep. 1946
8	AAA/132	Havildar	Parma Nand	N.K.		17 Sep. 1946
9	AAA/9097	Havildar	Prakasm Swaminatman	N.K.		17 Sep. 1946
10	AAA/1878	Havildar	Jagat Ram	N.K.		9 May 1946
11	AAA/2016`	Havildar	Sardar Khan	N.K.		9 May 1946
12	AAA/1744	Havildar	Rashan Din	N.K.		17 Sep. 1946

13	AAA/8301	Havildar	Ujagar Singh	N.K.	Burma	5 April 1945
14	AAA/8802	L/Havildar	Ghulam Mohd		Burma	5 April 1945
15	AAA/8338	L/Havildar	Ranjit Singh		Burma	5 April 1945
16	AAA/8356	L/Havildar	Gurabaksh Singh		Burma	5 April 1945
17	AAA/8432	L/Naik	Sadhu Singh		Burma	5 April 1945
18	AAA/8889	Gunner	Mokand Singh		Burma	5 April 1945
19	AAA/1774	RQMH	Gaz Singh		Burma	27 Sep. 1945
20	AAA/9242	Driver	Amar Singh		Burma	27 Sep. 1945

Note: N.K.: Not known.

Source: The details have been compiled from the notices published in *The London Gazette* and the records maintained in the *Regimental War Diaries*.

APPENDIX 8

Disbandment of AA Units

Unit	*Date of Disbandment*	*Place of Disband-ment*	*Remarks*
7 Sikh LAA Regt	30 September 1943	Mathura	Less 22 LAA Bty
14 Ind Lt AA Regt	31 October 1944	Mehgaon	
15 (Pb) Lt AA/AT Regt	7 November 1944	Imphal	
2 AATC Depot Bty	1 December 1944	Mehgaon	
26/9 (Pb) Hy AA Bty	15 February 1945	Mehgaon	
117 AA Ops Room	1 August 1945	Ranchi	
27/8 Sikh Lt AA Bty	4 August 1945	Ranchi	
102 Ind Lt AA Bty	4 August 1945	Ranchi	
102 AA Ops Room	4 August 1945	Ranchi	
VCO & NCO Trg Bty, AATC	6 August 1945	Mehgaon	
502 Ind Hy AA Bty	12 August 1945	Mehgaon	
41 HAA Bty, RIA	1 November 1945	Kamareddi	
13 HAA Bty, RIA	19 December 1945	Kamareddi	
14 HAA Bty, RIA	19 December 1945	Kamareddi	
10 HA Bty, RIA	28 December 1945	Mehgaon	
12 LAA Regt, RIA	15 January 1946	Mathura	
5 L.A.H.A. Regt, RIA	31 March 1946	Mathura	
17 LAA Regt, RIA	31 March 1946	Mathura	
8 Rajput HAA Regt	31 March 1946	Mathura	
8 Rajput HAA Bty	31 March 1946	Mathura	
9 LA Bty, RIA	15 May 1946	Mathura	
49 LAA Bty, RIA	15 May 1946	Mathura	
11 LAA Regt, RIA	30 June 1946	Mathura	
14 LAA Regt, RIA	30 June 1946	Mathura	
15 LAA Regt, RIA	15 July 1946	Mathura	
10 LAA Regt, RIA	31 July 1946	Mathura	

16 LAA Regt, RIA	31 July 1946	Coimbatore
29 LAA Regt, RIA	31 August 1947	Mathura
30 LAA Regt, RIA	31 August 1947	Mathura
31 LAA Regt, RIA	31 August 1947	Mathura
28 Para LAA Regt	30 September 1947	Mathura
19 HAA Regt, RIA	30 September 1947	Mathura

Source: Statistical Review of Personnel, Indian Army, vol. I to IV: 1939-1948, compiled by Army Statistical Organisation (A.S.O./2), published by Manager, Government of India Press, New Delhi.

Bibliography

Books Journals

Ahmad, Mustasad, *Living up to Heritage: History of Rajput Regiment*, Spantech & Lancer, New Delhi, 1997.

Ashmore, Major General E.B., *Air Defence*, Longmans Green & Co., London, 1929.

Banham, Tony, *Not the Slightest Chance: Defence of Hong Kong 1941*, Hong Kong University Press, Hong Kong, 2003.

Barkawi, Tarak, *Soldiers of Empire*, Cambridge University Press, Cambridge, 2017.

Birdwood, Colonel F.T., *Sikh Regiment in the Second World War*, Naval & Military Press, Uckfield, 2017.

Blackburn, Kevin, *Did Singapore have to Fall?* Routledge Curzon, London, 2005.

Brian, Farrell, *The Defence and Fall of Singapore*, Monsoon Books, Singapore, 2002.

Burton, John, *Fortnight of Infamy: The Collapse of Allied Airpower West of Pearl Harbour*, Naval Institute Press, Washington D.C., 2006.

Chhina, T.S. Rana, *The Indian Distinguished Service Medal*, Invicta India, New Delhi, 2001.

Crabtree, James T., *On Air Defence*, Praeger, London, 1994.

Doulton, A.J.F., *The Fighting Cock: Being the History of the 23rd Indian Division*, Naval & Military Press, Uckfield, 2002.

Evans, Bryn, *Air Battle for Burma: Allied Pilots' Fight for Supremacy*, Pen & Sword Aviation, Barnsley, 2016.

Farndale, Sir Martin, *History of the Royal Regiment of Artillery*, vol. VI: *The Far East Theatre 1941-46*, Brassey's, London, 2000.

Farrell, Brian, *The Defence and Fall of Singapore*, Monsoon Books, Singapore, 2017.

Frank, McLynn, *The Burma Campaign: Disaster into Triumph 1942-45*, Vintage Books, London, 2010.

Ganders, Terry, *The Bofors Gun*, Pen & Sword, Barnsley, 2013.

Grehan, John and Martin Mace, *Disaster in the Far East 1941-1942: The Defence of Malaya, Japanese Capture of Hong Kong, and the Fall of Singapore (Despatches from the Front)* , Pen & Sword, Barnsley, 2014.

Gruhl, Werner, *Imperial Japan's World War Two: 1931-1945,* Transaction Publishers, New Brunswick, 2013.

Gulati, Brigadier (Retd) Y.B., *History of the Regiment of Artillery: Indian Army,* Palit & Dutt Publishers, Dehradun, 1971.

Hall, Timothy, *The Fall of Singapore 1942*, Griffin Paperbacks, Sydney, 1992.

History of Army Air Defence, Published by Army Air Defence Association on behalf of the Directorate of Army Air Defence, Indian Army, 2008 (Restricted circulation).

Hodges, CB, CBE, Maj Gen B.P., *History of Royal Regiment of Artillery between the Wars 1919-39,* Brassey's, London,1991.

Hogg, Ian V., *Anti-Aircraft Artillery,* Crowood Press, Marlborough, 2002.

Hughes, David, *The British Armies in World War Two: An Organisational History,* vol. IX: *The Indian Army,* pt. 2: *The Indian Army in the East, 1939-43,* The Nafziger Collection, Inc, West Chester, Ohio, 2006.

______, *British Armies in World War Two: An Organizational History,* vol. X: *The Indian Army,* pt. 3: *The Indian Army in the East 1944-45,* The Nafziger Collection, Inc, West Chester, Ohio, 2006.

Johnson, Rob (ed.), *The British Indian Army: Virtue and Necessity,* Cambridge Scholars Publishing, Newcastle upon Tyne, 2014.

Katoch, Hemant Singh, *Imphal 1944: The Japanese Invasion of India,* Bloomsbury, London, 2018.

Kempton, Chris, *A Register of Titles of the Units of the H.E.I.C. & Indian Armies, 1666-1947,* British Empire and Commonwealth Museum, Bristol, 1997.

Khan, Yasmin, *The Raj at War: A People's History of India's Second World War,* Bodley Head, London, 2015.

Kirby, S. Woodburn, *The War Against Japan: The Decisive Battles,* vol. III, Natraj Publishers, Dehradun, 2002.

______, *The War Against Japan,* vol. IV: *The Reconquest of Burma,* Natraj Publishers, Dehradun, 2002.

Kreis, Jon F, *Air Warfare and Air Base Defence,* Office of Air Force History, United States Air Force Washington D.C., 1988.

Kwang, Chi Man, *Eastern Fortress: A Military History of Hong Kong 1840-1970,* Hong Kong University Press, Hong Kong, 2015.

Latimer, Jon, *Burma: The Forgotten War*, John Murray, London, 2004.

Lockwood, David, *Calcutta Under Fire*, Rupa, New Delhi, 2019.

Lyman, Robert, *Japan's Last Bid for Victory: The Invasion of India, 1944*, Pen and Sword, Barnsley, 2011.

Maurice-Jones, Colonel K.W., *The History of Coast Artillery in the British Army*, Royal Artillery Institution, London, 1959.

Moreman, Tim, *The Jungle, Japanese and the British Commonwealth Armies at War, 1941-45*, Taylor and Francis, London, 2005.

Owen, Frank, *The Fall of Singapore*, Michael Joseph Ltd, London, 1960.

Palit, Maj Gen D.K., *History of Regiment of Artillery Indian Army*, Lee Cooper, London, 1972.

Pearson, Michael, *The Burma Air Campaign: 1941-1945*, Pen & Sword, Barnsley, 2002.

Percival, Lt General A.E., *War in Malaya*, Eyre & Spottiswoode, London, 1949.

Phythian-Adams, Lt Col EG, OBE, *The Madras Soldier, 1746-1946*, Government Press, Madras, 1948.

Prasad, Bisheshwar (ed.), *Campaigns in South-East Asia 1941-42*, Pentagon Press, New Delhi, 2012.

______, *The Retreat from Burma 1941-42*, Pentagon Press, New Delhi, 2014.

______, *The Reconquest of Burma*, vol. I: *1942*, Pentagon Press, New Delhi, 2014.

______, *The Reconquest of Burma*, vol. II: *June 1944-June 1945*, Pentagon Press, New Delhi, 2014.

______, *The Campaign in Italy 1943-45*, Pentagon Press, New Delhi, 2014.

______, *Expansion of the Armed Forces and Defence Organisation 1939-45*, Pentagon Press, New Delhi, 2014.

Preston-Hough, Peter, *Commanding Far Eastern Skies*, Helion and Company, Warwick, 2019.

Prestwood Lucas, Sir Charles, *The Empire at War*, vol. 15, Oxford University Press, London, 1926.

Qureshi, Ibrahim, *The First Punjabis: History of the First Punjab Regiment*, Gale & Polden, London, 1958.

Reed, Sir Stanley, *India Year Book*, vol. XXII, Bennett, Coleman and Co., Bombay, 1936.

Rose, Patrick, *The Indian Army, 1939-47: Experience and Development*, Routledge, London, 2012.

Rottman, Gordon, *Japanese Army in World War II: Conquest of the Pacific 1941-42*, Osprey Publishing, Oxford, 2005.

Routledge, R.N., *History of the Royal Regiment of Artillery Anti-Aircraft Artillery 1914-1955*, Brassey's, London, 1994.

Roy, Kaushik, *The Indian Army in the Two World Wars*, Brill, Singapore, 2012.

———, *Sepoys Against the Rising Sun: The Indian Army in the Far East and South-East Asia 1941-45*, Brill, Singapore, 2016.

———, *Tropical Warfare in the Asia-Pacific Region, 1941-45*, Routledge, London, 2018.

Slim, William, *Defeat into Victory*, Cooper Square Press, New Delhi, 2002.

Smith, Colin, *Singapore Burning: Heroism and Surrender in World War II*, Penguin, London, 2002.

Statistical Review of Personnel, Army of India, Compiled by Army Statistical Organisation, vols. 1 to 4, Held with History Division, Ministry of Defence, New Delhi.

Stille, Mark, *Malaya and Singapore 1941-42: The Fall of Britain's Empire in the East*, Osprey Publishing, Oxford, 2016.

Talwar, Sushil, *Indian Recipients of the Military Cross* (vol. I & II), KW Publishers, New Delhi, 2017.

The Asiatic Journal and Monthly Register for British and Foreign India, China and Australasia, vol. 38, 1 January 1842.

Thompson, Julian, *Forgotten Voices of Burma*, in Association with The Imperial War Museum, Ebury Press, London, 2009.

Tsuji, Colonel Masanobu, *Singapore: The Japanese Version*, Constable and Company, London, 1960.

Warren, Ala, *Burma 1942: The Road from Rangoon to Mandalay*, Bloomsbury, London, 2012.

Warwick, Nigel N.M., *Constant Vigil: The RAF Regiment in the Burma Campaign*, Pen and Sword Aviation, Barnsley, 2007.

War Despatches

'Air Operations in Burma and Bay of Bengal from 1 January to 22 May 1942', Supplement to *The London Gazette* of Tuesday, the 5 March 1948.

'Air Operations in South East Asia from 6 November 1943 to 31 May 1944', Supplement to *The London Gazette* of Tuesday, the 31 March 1951.

'Air Operations in South East Asia from 1 June 1944 to the occupation of Rangoon 2 May 1945', Supplement to *The London Gazette* of Friday, the 5 April 1951.

'Operation in Assam and Burma from 23 June 1944 to 12 November 1944', Supplement to *The London Gazette* of Friday, the 6 April 1951.

'Operation in Burma from 12 November 1944 to 15 August 1945', Supplement to *The London Gazette* of Friday, the 6 April 1951.

'Operation in Eastern Theatre, based in India, from March 1942 to 31 December 1942', Supplement to *The London Gazette* of Tuesday, the 17 September 1946.

Operations in Hong Kong from 8 December 1941 to 25 December 1941', Supplement to *The London Gazette* of Thursday, the 29 of January, 1948.

'Operations of Malaya Command from 8 December 1941 to 15 February 1942', Supplement to *The London Gazette* of Friday, the 20 February 1948.

'Operation in the India Command from 1 January 1943 to 15 November 1943', Supplement to *The London Gazette* of Tuesday, the 27 April 1948.

'Operation in the Indo-Burma Theatre, based in India from 21 June 1943 to 30 June 1943', Supplement to *The London Gazette* of Tuesday, the 20 April 1948.

Report on the Air Operations during the Campaigns in Malaya and Netherlands East Indies from 8 December 1941 to 12 March 1942', Supplement to *The London Gazette* of Friday, the 20 February 1948.

Regimental Histories

"Tiger Head: The Story of 26th Indian Division", Held with Pyare Lal Library, United Service Institution of India, New Delhi.

Army Lists (Published by Army Department, Government of India, Superintendent, Government Printing Press, Calcutta).

Indian Army List for July 1941.
Indian Army List for April 1943.
Indian Army List for October 1943.
Indian Army List for April 1945 .

War Diaries (History Division, Ministry of Defence, Government of India)

File No 601/2077/WD	Headquarters Eastern Army (AA Branch)
File No 601/2098/WD	Headquarters 1 Indian AA Brigade
File No 601/2099/WD	Headquarters 2 Indian AA Brigade
File No 601/2100/WD	Headquarters 3 Indian AA Brigade
File No 601/2019/WD	Headquarters 4 AA Brigade
File No 601/2093/WD	Headquarters 9 AA Brigade, RA
File No 601/2094/WD	Headquarters 13 AA Brigade, RA
File No 601/2095/WD	Headquarters 23 AA Brigade, RA
File No 601/2096/WD	Headquarters 24 AA Brigade, RA
File No 601/2103/WD	1 R.M. AA Brigade
File No 601/2290/WD	1 Indian HAA Regiment
File No 601/2281/WD	1 HAA Regiment, RA (B.A.F.)
File No 601/2302/WD	7 (Rajput) HAA Regiment
File No 601/2316/WD	25 Indian HAA Regiment
File No 601/2313/WD	Reserve AA Regiment
File No 601/2314/WD	Reserve Mobile AA Regiment
File No 601/2348/WD	1 Indian HAA Battery
File No 601/2427/WD	6 (Punjab) LAA Regiment (later 28 (Punjab) Para LAA Regiment)
File No 601/2431/WD	7 (Sikh) LAA Regiment
File No 601/2436/WD	1 Indian LAA Regiment (Now 25 LAA Regiment)
File No 601/2437/WD	2 Indian LAA Regiment
File No 601/2438/WD	3 Indian LAA Regiment (Now 27 LAA Regiment)
File No 601/2439/WD	4 Indian LAA Regiment (Now 29 LAA Regiment)
File No 601/2./WD	11 Indian LAA Regiment
File No 601/2497/WD	3 Indian LAA Battery
File No 601/2510/WD	102 Indian LAA Battery (Later 102 Composite AA Battery)
File No 601/2511/WD	103 Indian LAA Battery (Later 103 Composite AA Battery)
File No 601/2518/WD	5 Indian LAA Battery
File No 601/2694/WD	Army AA School
File No 601/2695/WD	AA Training Establishment, Ceylon
File No 601/2709/WD	No 4 AATC WAC(I)

War Diaries (National Archives, Kew (UK)

WO 172-196	1 Indian HAA Regiment Aug-Oct, Dec 1941
WO 172-630	1 Indian AA Brigade July Nov 1942
WO 172-631	2 Indian AA Brigade June-Dec 1942
WO 172- 2447	2 Indian LAA Regiment Jan-Feb, Dec 1943
WO 172-2454	3 Indian LAA Battery January- May, July-Dec 1943
WO 172-4468	3 Indian AA Brigade Apr-Dec 1944
WO 172-14955	25 Indian HAA Regiment Mar-Dec 1943

Films with Imperial War Museum

JFU-261	Allied Victory Parade at Rangoon in June 1945
JFU-268	Allied Victory Parade at Rangoon in June 1945

Oral History (Imperial War Museum)

Hoskin, Ernie, 6th Battery/1st Indian LAA Regiment accessed at Imperial War Museum at https://www.iwm.org.uk/collections/item/object/80027023 on 13 June 2019.

Lancelot Lester Nash, 3rd Indian LAA Battery accessed at Imperial War Museum https://www.iwm.org.uk/collections/item/object/80020144 on 10 June 2019.

MacFetridge, C.H.T., 3rd Indian LAA Battery accessed at Imperial War Museum at https://www.iwm.org.uk/collections/item/object/80020185 on 16 June 2019.

Index

14th West African AA Brigade 126, 129, 138, 144, 149
18 pounder Mk I gun 36, 93
3-inch 20 cwt AA gun 19
AA Guns in Ground Role 118, 128, 145-7, 148, 152-3, 186-7, 202
AA School 131, 178-9, 188
Abit Ali, Jemadar 89, 223
Addu Atoll 129, 160, 162-3
Afghanistan 18, 23-4, 93
Afzal, Lieutenant Colonel Mohammad 11
Agartala 215-16
Aircraft recognition 180
Air Ministry Experimental Stations (AMES) 103-4, 122
Air raid warning 61, 114
Akyab Island 132, 135, 138-9, 148-50, 220
Allied Air Forces 58, 63, 112, 120, 142-4
Allied Land Forces South East Asia (ALFSEA) 142, 149, 161
Alexandra 57, 62, 64, 67-8
Alor Star 43, 56
Ambala 32
American Volunteer Group (AVG) 74, 82
Amphibious Operations 149, 164
Anti-Dacoity operations 168
Arakan 103, 106, 116, 120-1, 126, 128, 132, 137-9, 141, 144, 147, 161, 203, 218
Asansol 102, 104, 215, 217
Asiatic Artillery 40
Assam 86, 93, 96-9, 102-8, 114, 117, 119, 121, 126, 138, 141-5, 149, 155, 161, 167, 184, 215-19
Auchinleck 22

Balbir Singh, Gunner 68, 187, 222
Bangalore 26, 209
Barambe 127
Basra 158, 210
Battle of Admin Box 140-1
Blakang Mati 55, 57
Bombay 30, 34, 38, 77, 97, 101-6, 149, 158, 163, 165, 167, 178, 205, 215, 218
British Army
- 14th Army 69, 136, 144-7, 150-1, 160
- 18th Division 67
- 7th Armoured Brigade Group 82

British Expeditionary Force 18
British NCOs 31, 35, 98, 118, 130, 178
British officers 36, 53-4, 57, 131, 164
Burma Auxiliary Force (BAF) 13, 26, 32, 72, 74-5, 90
- 1st HAA Regiment, RA 72
- 1st HAA Battery, RA 74
- 3rd LAA Battery 74-5

Burma Corps 82-3, 85, 87, 91
Burma Rifles 74

Calcutta 96, 101-6, 108, 110, 116-17, 119, 122, 168, 178, 207-9, 215-17, 219-20
Ceylon 40, 99, 103, 145, 159-63
Ceylon Garrison Artillery 159
Chabua 107-8, 145
Chak Abu 37
Chaklala 172
Chain Overseas Low (COL) 75
Chatfield Committee 21-2, 26
China Gun Lascars 39-40
Chindits 133, 135, 137, 141
Chindwin 85-7, 89, 126, 137, 141, 144-7
Chittagong 103, 106-7, 116, 119, 138, 215-19
Cochin 103
Colaba 34, 38, 105
Colombo 149, 159-62
Committee of Imperial Defence 25
Control and Reporting System 97, 104, 114, 164

Defence of India plan 18
Deolali 35, 96, 99, 131-2, 164, 206, 208
Digboi 35-6, 93, 95, 98, 102, 104, 106, 113-14, 117, 119, 146, 208, 215-16, 218-19
Dimapur 97, 102, 119, 141, 143-5, 149, 215-19
Dinjan 107-8, 113, 115, 119, 145
Direct enlistment of VCOs 31, 130
Distinguished Service Order (DSO) 68, 177, 187, 221
Drigh Road 36, 206
Dummies 160
Dundeedaw 74

Early Warning (EW) 95-7, 104, 111, 114-15, 159
  EW Radar 96, 104, 159
Eastern Army School for Mobile AA Artillery 105
Electronic Warfare (EW) 160, 184
External Defence Troops 22

Far East Command 72
Fratricide 77

Garran Tribunal 20
Germany 19, 25-6, 160
GL set 64, 77, 95, 115, 159, 162
Gobindpur Hill 104
Gulf of Martaban 75
Gulunche 140, 164, 171, 211
Gun Night 77
Gun Operations Room (GOR) 50, 96-7, 104, 207-8
Gun Position Officer (GPO) 112

Habbaniya 157-8
Ha-Go counter-offensive 139
Hari Singh, Gunner 121, 222
Hispano 20 mm cannon 104
Hlegu 82
HMS Prince of Wales 63
HMS Repulse 63
Hopson, Lt John Ashcroft 67, 71, 221
Hong Kong Volunteer Defence Corps 41, 50

Imperial Japanese Army
  18th Division 67
  31st Division 143
  33rd Division 143
  38th Division 50
  5th Division 67
  54th Division 139
  55th Division 78, 139
Imperial Japanese Army Air Force
  3rd Air Corps 58
  3rd Air Division 49, 183
  5th Air Corps 58
  5th Air Division 84, 183
  Air Combat Groups 58
  Aircraft 58, 60, 63, 65, 78, 81, 87

Imperial Japanese Navy 25, 58, 60, 183
Imperial Japanese Navy Air Arm
22nd Air Flotilla 58, 183, 193
Genzan Air Group 61
Kanoya Air Group 58
Mihoro Air Group 601-61
Type 96 Mitsubishi A5M4 Claude fighter 58
Type 96 Mitsubishi G3M1 Nell bomber 58, 60
Imphal 91, 99, 103, 116, 123, 129, 137, 141-6, 149, 202, 204, 218-19, 225
Indian Air Force 110, 189
No 28 Squadron, IAF 11
Indian Anti-Aircraft Artillery
Anti-Aircraft (AA) Brigades 94, 96, 102-5, 109, 111-16, 130, 215
1 AA Brigade 96, 102, 105, 171, 207, 215-19
2 AA Brigade 104, 116, 171, 208, 215-19
3 AA Brigade 105, 145-6, 149, 151-4, 171, 181, 208, 215-19
Anti-Aircraft Operations Room (AAOR) 161, 163, 171, 207-8
Anti-Aircraft Training Centre
No 1 AATC 96, 98, 132, 158, 205
No 2 AATC 96, 98, 132, 183, 225
No 3 AATC 98, 105, 132, 208
No 4 AATC 116-17, 209, 232
Composite AA batteries 160, 162, 232
Heavy Anti-Aircraft (HAA) Batteries
1 HAA Battery 177, 212
19 HAA Battery 212, 216
2 HAA Battery 55
20th HAA Battery 43, 111
3 HAA Battery 62, 67
4 HAA Battery 43, 212
8 HAA Battery 77, 80, 87, 171
Light Anti-Aircraft (LAA) Batteries
19 LAA Battery 173
3 LAA Battery 75, 79-81, 85, 161
Heavy Anti-Aircraft (HAA) Regiment
1 HAA Regiment 30, 34, 36, 37, 43, 59, 64, 67-9, 74, 76, 91, 95, 176-7, 180
10 HAA Regiment 170, 182, 213
11 HAA Regiment 170, 209, 213
12 HAA Regiment 170, 210, 213
13 HAA Regiment 171, 210, 213
14 HAA Regiment 171, 210, 213
15 HAA Regiment 211, 213
16 HAA Regiment 211
18 HAA Regiment 206
2 HAA Regiment 36-7, 212
20 HAA Regiment 171, 209, 213
21 HAA Regiment 171, 210, 213
25 HAA Regiment 148, 210, 213, 215
3 HAA Regiment 170, 212
4 HAA Regiment 37, 104, 209
5 HAA Regiment 146, 170, 182, 206-7, 212
6 HAA Regiment 148
7th Rajput HAA Regiment 97, 102, 110-11
8th Rajput HAA Regiment 97
9th Punjab HAA Regiment 97, 170, 182
'R' HAA Regiment 30, 34, 36-7
'S' HAA Regiment 36-7
Light Anti-Aircraft (LAA) Regiments
1 LAA Regiment 102, 106-8, 114, 151-2, 154-5, 168, 171, 182, 213
11th LAA Regiment 148, 209, 214
14th LAA Regiment 130, 160, 211, 214

17th LAA Regiment 118, 161, 170, 209, 214
2 LAA Regiment 37, 104, 142, 145-7, 151, 171, 202, 205-6
25 LAA Regiment 174, 190, 205
27 LAA Regiment 172, 174, 181-2, 190, 207, 232
28 (Punjab) Para LAA Regiment 172-3
29 LAA Regiment 172, 209, 226
3 LAA Regiment 96, 146, 161, 171, 185-6, 215
4 LAA Regiment 209, 213
6 (Punjab) LAA Regiment 213
7 (Sikh) LAA Regiment 213
8th Sikh LAA Regiment 97, 121
9th Rajputana LAA Regiment 163-5, 168
'U' LAA Regiment 35
Self-Propelled (SP) LAA Regiment 152
Indian Army
111th Indian Infantry Brigade 137
11th Army Group RIA 171
12th Army Group RIA 171
77th Indian Infantry Brigade 135
Corps
3 Corps 43
4 Corps 119, 128, 142-4, 146-7
33 Corps 129, 143-7, 149
Divisions
11th Indian Division 61
14th Indian division 132
17th Indian Division 74, 77-8, 80-2, 84-5, 88, 142, 147, 150-1
19th Indian Division 146-7
20th Indian Division 142, 146
23rd Indian Division 142-3, 165, 201
25th Indian Division 165
26th Indian Division 148-50
36th Indian Division 139
4th Indian Division 26
5th Indian Division 139-40, 145, 150, 165
7th Indian Division 138-40, 142
General Reserve 22-4, 128, 203
Indianisation 20, 23
Mechanisation 22
Indian Artillery
Coastal Defence & Anti-Aircraft School 30
Field Brigade 23
Field Artillery Training Centre (FATC) 33-4, 129, 158, 183
Mountain Artillery 23, 29, 31-2, 93, 176
Mountain Artillery Training Centre (MATC) 32-4
School of Artillery 34-5, 170
Cadet Wing 35
Indian Distinguished Service Medal (IDSM) 68, 90, 187, 222
Indian National Army (INA) 42, 45, 69, 182
Indian Observer Corps 52, 116
Control Units 116
Indian State Forces 53-4, 56
Irrawaddy River 84
Italy 25-6

Jamshedpur 101, 104-5, 161, 215, 217
Japanese air raids 60, 62-3, 66, 81, 88, 101, 106-8, 114, 118-21, 141, 144, 147, 150, 167
Johore 25, 43, 52, 59-60
Jorhat 112, 119, 145, 149, 161, 220

Kai Tak 48-50
Kakul 34-5
Kaladan 139, 147
Kalantan 57
Kalewa 85-9, 146, 153
Kallang 63, 67
Kanchrapara 104

Karachi 30, 34-8, 96, 131, 158, 174, 176, 178-9, 188, 201, 205-6
Kharagpur 104
Kohima 128, 141-5
Kota Bharu 43, 56, 58, 61
Kowloon 48
Kuantan 58
Kumbhigram 121

Land Mattress 148
Lashkar Singh, Lt 151
Lines of Communication 83, 128, 135, 203
Ledo Road 126, 137, 148, 216-18
Lentaigne 89, 137

MacFetridge, Major Charles HT 35-6, 75-6, 79, 85, 87, 89, 93, 95, 187
Madras Sappers and Miners 158
Madrassi units 30, 132
Magwe 168
Maldives 161-2
Malir Cantt 34-6, 75, 93, 95, 107, 131-2, 158, 179, 207, 209-11
Malta 17
Malaya Command 37, 52, 56, 95
Mandalay 84-5, 126, 137, 146-7
Manipur 91, 97, 103, 141, 145-7, 203
Martaban 75, 77-9
Mathura 33-4, 129, 158, 172, 183, 225-6
Mehar Dass , 2nd Lt 79-80, 181
Mehgaon 98, 105, 131-2, 158, 163, 208-11, 215
Mehnga Singh, Gunner 121, 213
Meiktila 82, 84, 147, 149-50, 220
Mergui 78
Mersing 52, 60
Military Cross (MC) 67-8, 81, 90, 187, 221
Military Engineering Service (MES) 102
Mingaladon 74-5, 81, 150-2
Mobile Role, use in 130, 154, 161, 183
Mohanbari 107, 145
Mohd Amir 44-5, 186
Morale 66, 79, 108-9, 112, 135, 137, 153, 167, 173
Mosul oilfields 19
Moulmein 75-81, 90
Mount Pleasant 44
Multi-Rocket Launching Systems (MLRS) 148
Myitkina 126, 137, 141, 146

Naval Base (East) 54
No Fly Zones 114
Normanton 57, 62, 67
North West Frontier Province 17
Nur Khan, Subedar 82, 225

Operation Longcloth 135
Operation Zipper 149, 163-5

Pakistan 173-4, 189-90, 205-6
Partition 172-3, 182, 206-7, 211
Passive Air Defence (PAD) 105
Pegu 80-1, 90, 150-1
Percival, Lt Gen AE 48, 66, 68
Peshawar 18, 20, 23, 208
Predictor 198-9
Prome 82-4, 90, 150
Poona 140, 163-4, 168, 171-2, 218

Radars 52, 60, 95-6, 99, 103-4, 115-16, 122, 144, 159-60, 162, 180, 184, 189-90
Raffles 57, 61
RAF Regiment Training Centre 104
Ramree Island 148-50, 220
Ranchi 127-8, 218, 225
Rangoon 72, 74-7, 79, 80, 82-3, 90-1, 132, 144, 147, 149-52, 167-8, 182, 187, 220
Rangoon Field Brigade 72, 91

Recruitment 20, 29-30, 32-3, 39, 176, 182
Re-organisation 40, 99, 125-7, 169-73, 203
Reorganisation Committee (India) 169
Royal Air Force 46, 49-50, 56, 58-61, 64, 77-8, 82, 103-4, 107, 112, 157-8, 162-3, 199
  Aircraft
    Blenheims 74, 78
    Brewster Buffaloes 49, 74
    Hawker Hurricane 64
    Wildebeestes 49-50
  Balloon Squadrons 100, 110
  RAF Regiments 103-4
Royal Artillery
  AA Brigades
    13 AA Brigade, RA 98, 106, 104-5, 144, 148, 215-16
    23rd AA Brigade 159-61
    24th AA Brigade 145, 149-51, 155, 159-61, 188, 220
    4th AA Brigade 157-8
    5th AA Brigade, RA 20
    9th AA Brigade, RA 107, 113, 118-19, 140, 149, 155, 163, 167, 215-18, 220
  Batteries
    8th AA Battery, RA 18-19, 23, 30, 32
    8th HAA Battery, RA 77, 84, 86-7, 176
  Regiment
    12th Coast Regiment, RA 50
    3rd AA Regiment, RA 52
    5th AA Regiment, RA 26, 40-4, 50
    69th LAA Regiment, RA 133, 141, 146
Royal Artillery Training Centre 33
Royal Australian Air Force (RAAF) 60
Royal Garrison Artillery (RGA) 17
Royal Indian Airforce 110
Royal Marines 159
Royal Navy 46, 52, 62, 162
Russia 18, 23, 93, 121, 158

Salbani 104
Sardar Khan, Havildar 89, 223
Scale of AA Defences 117-18, 129-30, 144-9
Searchlight 17, 46, 54, 60-1, 77, 101, 105, 161
Secunderabad 104, 163
Selatar 25, 57, 60-1
Sembawang 60-1, 63-4
Semi-mobile units 130, 154
Sham Lal, Havildar 68, 187, 222
Shwebo 85-6
Shwedaung 83, 168
Shwegyin 85-90
Singh, Amar 44, 224
Singh, Mohan 42
Singh, Udham 44-5
Sittang river 81-2, 84
Slim, Lt General William 85-6, 126, 137-8, 140-1, 144, 147
South East Asia Command (SEAC) 73, 125-6, 133, 137-8, 140, 147, 150, 163
Suda Rang, Gunner 89, 222
Sungei Patani 43, 56
Syriam 72, 74, 77, 153

Taukkyan 82-3, 85
Technical Training Battery 30, 34, 38, 205
Tenasserim 75, 78
Tengah 43, 61, 63-4, 66, 187
Thailand 58, 75, 78
Tharrawaddy 83
Tinsukhia 114-15, 119, 146, 215-16, 218
Training 65, 93, 95-6, 98-9, 101, 103-5, 116-18, 125-7, 130-2

Training Centres 32-4, 163, 177, 180, 183
Trincomalee 99-100, 103, 159-61
Toungoo 82, 150-2, 220

United States 49, 129
United States Anti-Aircraft Artillery (AAA) 129
Urdu, use of 31-2, 35-7, 54, 164, 178, 180-1
US Army Air Force 112, 129

Viceroy Commissioned Officers (VCO) 30-1, 34, 53, 85, 89, 130, 158, 168, 176, 181, 225
Vickers 2-pounders 52
Victory parade 151, 154
Vizag 100-1, 103, 116, 216-17

Wapitis 105
War Establishment (WE) 103
Williamson, Lt Colonel John Rowley 68, 221
Women Auxiliary Corps (India) 99, 116-17, 209

Yenangyaung 84-5, 150

Zayatkwin 74, 151